T0173124

WJEC
Mathematics
for A2 Level – Pure

Stephen Doyle

Illuminate Publishing

Published in 2018 by Illuminate Publishing Ltd, P.O Box 1160, Cheltenham, Gloucestershire GL50 9RW

Orders: Please visit www.illuminatepublishing.com
or email sales@illuminatepublishing.com

British Library Cataloguing in Publication Data

A catalogue record for this book is available from the British Library

ISBN 978-1-911208-54-9

Printed by Cambrian Printers, Aberystwyth

06.19

This material has been endorsed by WJEC and offers high quality support for the delivery of WJEC qualifications. While this material has been through a WJEC quality assurance process, all responsibility for the content remains with the publisher.

The publisher's policy is to use papers that are natural, renewable and recyclable products made from wood grown in sustainable forests. The logging and manufacturing processes are expected to conform
to the environmental regulations of the country of origin.

Editor: Geoff Tuttle
Cover design: Neil Sutton
Text design and layout: GreenGate Publishing Services, Tonbridge, Kent

Photo credits

Cover: Klavdiya Krinichnaya/Shutterstock; **p9** Gajus/Shutterstock; **p17** Maren Winter/Shutterstock; **p52** gorosan/Shutterstock; **p74** Yury Zap/ Shutterstock; **p103** Chad Bontrager/Shutterstock; **p136** Jakkrit Orrasri/ Shutterstock; **p149** Kirill Neiezhmakov/Shutterstock; **p175** Mvolodymyr/ Shutterstock.

Acknowledgements

The author and publisher wish to thank Sam Hartburn and Siok Barham for their careful attention when reviewing this book.

Contents

Contents

How to use this book

The contents of this study and revision guide are designed to guide you through to success in the Pure Mathematics component of the WJEC Mathematics for A2 level: Pure examination. It has been written by an experienced author and teacher and edited by a senior subject expert. This book has been written specifically for the WJEC A2 course you are taking and includes everything you need to know to perform well in your exams.

Knowledge and Understanding

Topics start with a short list of the material covered in the topic and each topic will give the underpinning knowledge and skills you need to perform well in your exams.

The knowledge section is kept fairly short, leaving plenty of space for detailed explanation of examples. Pointers will be given to the theory, examples and questions that will help you understand the thinking behind the steps. You will also be given detailed advice when it is needed.

The following features are included in the knowledge and understanding sections:

- **Grade boosts** are tips to help you achieve your best grade by avoiding certain pitfalls which can let students down.

- **Step by steps** are included to help you answer questions that do not guide you bit by bit towards the final answer (called unstructured questions). In the past you would be guided to the final answer by the question being structured. For example, there may have been parts (a), (b), (c) and (d). Now you can get questions which ask you to go to the answer to part (d) on your own. You have to work out for yourself the steps (a), (b) and (c) you would need to take to arrive at the final answer. The 'step by steps' help teach you to look carefully at the question to analyse what steps need to be completed in order to arrive at the answer.

- **Active learning** – are short tasks which you carry out on your own which aid understanding of a topic or help with revision.

- **Summaries** – are provided for each topic and present the formulae and the main points in a topic. They can be used for quick reference or help with your revision.

Exam Practice and Technique

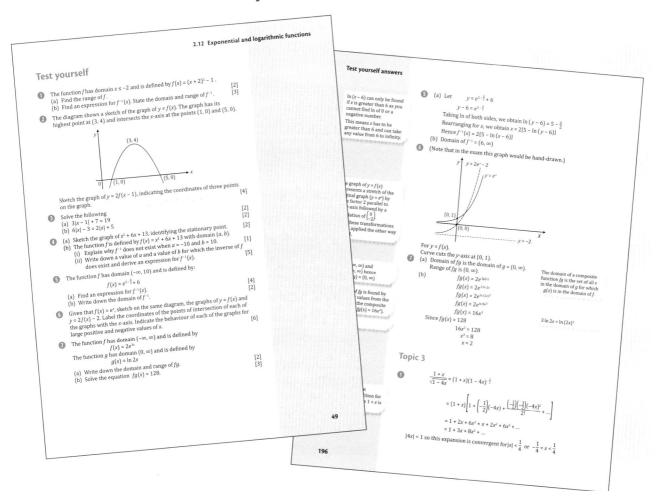

Helping you understand how to answer examination questions lies at the heart of this book. This means that we have included questions throughout the book that will build up your skills and knowledge until you are at a stage to answer full exam questions on your own. Examples are included, some of which are full examination style questions. These are annotated with pointers and general advice about the knowledge, skills and techniques needed to answer them.

There is a Test yourself section where you are encouraged to answer questions on the topic and then compare your answers with the ones given at the back of the book. There are many examination standard questions in each test yourself that provide questions with commentary so you can see how the question should be answered.

You should, of course, work through complete examination papers as part of your revision process.

We advise that you look at the WJEC website www.wjec.co.uk where you can download materials such as the specification and past papers to help you with your studies. From this website you will be able to download the formula booklet that you will use in your examinations. You will also find specimen papers and mark schemes on the site.

WJEC Mathematics For A2 Level Pure & Applied Practice Tests

There is another book which can be used alongside this book. This book provides extra testing on each topic and provides some exam style test papers for you to try. I would strongly recommend that you get a copy of this and use it alongside this book.

Good luck with your revision.

Stephen Doyle

1 Proof

Introduction

You came across proofs in Topic 1 of the AS Pure Unit 1 course. A mathematical proof is a mathematical argument that convinces others that a mathematical statement is true. In this topic you come across one more technique to prove or disprove mathematical statements. This topic is concerned with a mathematical proof called proof by contradiction. As with all proofs, proof by contradiction depends on precise use of English as well as mathematical techniques.

This topic covers the following:

1.1 Real, imaginary, rational and irrational numbers

1.2 Proof by contradiction

1.3 Proof of the irrationality of $\sqrt{2}$

1.4 Proof of the infinity of primes

1.5 Application of proof by contradiction to unfamiliar proofs

1.1 Real, imaginary, rational and irrational numbers

You came across this material in AS Pure, but as we will be using the terminology a lot in this topic, here is a reminder:

Real numbers are the numbers we are used to using. So 2, 0, 0.98, $\frac{1}{3}$, −3, π, $\sqrt{2}$, etc., are all examples of real numbers.

Imaginary numbers are those numbers that when squared give a negative number. So $\sqrt{-1}$ is an imaginary number as $(\sqrt{-1})^2 = -1$.

Rational numbers can be expressed as fractions $\left(\text{i.e. } \frac{a}{b}, \text{ where } b \neq 0\right)$.

Irrational numbers cannot be expressed as fractions. Numbers where there are numbers after the decimal point that go on forever and do not repeat are irrational, e.g. $\sqrt{2}$ and π.

1.2 Proof by contradiction

BOOST

Grade ⇧⇧⇧⇧

Proofs are difficult because they can apply to so many parts of the specification. This is an area where you really need lots of practice before you become proficient. Stick with it.

Proof is concerned with the demonstration of the truth of a conjecture. The essential steps in proof by contradiction are to **assume** that the conjecture is **false** and then show that this assumption leads to a **contradiction**.

Sometimes it is difficult or almost impossible to prove that a conjecture such as '$\sqrt{2}$ is irrational' is true. The problem is that there are an infinite number of numbers whose squares might equal 2, so we could not test them all. It is easier to consider the contradiction which is that '$\sqrt{2}$ is rational'. We than have to consider logical steps based on what we know is mathematically true and when we meet something we know is mathematically false it means that the assumption made for the contradiction is incorrect and that the original assumption was indeed true.

Examples

1 Complete the following proof by contradiction to show that if n is an integer such that n^2 is divisible by 3, then n is divisible by 3.

Assume that n does not have a factor 3 so that

$$n = 3k + r$$

where k and r are integers and $0 < r < 3$.

$r = 1$ or 2

· ·

Answer

1 Then $n^2 = (3k + r)^2$
$$= 9k^2 + 6kr + r^2$$

which does not have a factor of 3.

$r^2 = 1$ or 4

But n^2 has a factor of 3 (given).

∴ there is a contradiction and our assumption that n does not have a factor of 3 is false.

2 Prove by contradiction the following proposition.

When x is real and positive, $4x + \dfrac{9}{x} \geq 12$.

The first line of the proof is given below.

Assume that there is a positive and real value of x such that $4x + \dfrac{9}{x} < 12$

Answer

2
$$4x + \frac{9}{x} < 12$$

Multiplying both sides by x gives
$$4x^2 + 9 < 12x$$

Subtracting $12x$ from both sides gives
$$4x^2 - 12x + 9 < 0$$

Hence $\quad (2x - 3)(2x - 3) < 0$

So $\qquad (2x - 3)^2 < 0$

For $(2x - 3)^2$ to be negative (i.e. < 0), $2x - 3$ would have to be an imaginary number. This contradicts that $2x - 3$ and x are real.

Hence $\qquad 4x + \dfrac{9}{x} \geq 12$

> Note we can multiply both sides by x without worrying about reversal of the inequality as x is a positive number.

> The left-hand side of the inequality is a quadratic function and can therefore be factorised.

> For any real and positive value of x, $(2x - 3)^2$ will always be greater than or equal to zero. It can never be negative if $2x - 3$ is real.

1.3 Proof of the irrationality of √2

The Greeks discovered that the diagonal of a square of side 1 unit has a diagonal whose length is not rational $\left(\text{i.e. it cannot be expressed as the fraction } \frac{a}{b}\right)$.

They used Pythagoras' theorem to find that the length of the diagonal was $\sqrt{2}$ so this meant $\sqrt{2}$ was irrational.

Using proof by contradiction to prove the irrationality of √2

We start off by supposing that $\sqrt{2}$ is rational and can therefore be expressed as a fraction in the form $\frac{a}{b}$.

So, we have $\sqrt{2} = \frac{a}{b}$ where a and b do not have any common factors.

Squaring both sides, we obtain $2 = \dfrac{a^2}{b^2}$

Rearranging this equation gives $2b^2 = a^2$ and this means a^2 is even as it has a factor of 2.

For a^2 to be even, a has to be even because if you square an even number you always obtain an even number and if you square an odd number you always get an odd number.

If a is even then a^2 must be divisible by 4. This means that b^2 and hence b must be even.

Now if a and b are both even, this is a contradiction to the assumption that a and b do not have any common factors.

Hence $\sqrt{2}$ cannot be rational so it is therefore irrational.

BOOST

Grade ⇧⇧⇧⇧

It is not easy to think up the proof for the irrationality of surds such as $\sqrt{3}$ or the infinity of primes. It is therefore worth spending time memorising them.

Examples

1 Complete the following proof by contradiction to show that $\sqrt{3}$ is irrational.

Assume that $\sqrt{3}$ is rational. Then $\sqrt{3}$ may be written in the form $\frac{a}{b}$ where a and b are integers having no common factors.

$\therefore a^2 = 3b^2$

$\therefore a^2$ has a factor 3.

$\therefore a^2$ has a factor 3 so that $a = 3k$ where k is an integer.

> Note that the symbol $\therefore$ means therefore.

Here we substitute $a = 3k$, into $a^2 = 3b^2$.

a and b should have no common factors.

This is because a and b were assumed to have no common factors and we have proved the existence of a common factor.

BOOST

Grade ⬆⬆⬆⬆

Make sure that you mention the word **contradiction** and say exactly what the contradiction is.

Answer

1       $\therefore$ $(3k)^2 = 3b^2$

       $\therefore$ $9k^2 = 3b^2$

       $\therefore$ $3k^2 = b^2$

       $\therefore b^2$ has a factor 3

       $\therefore b$ has a factor 3

a and b have a common factor (i.e. 3)

This is a contradiction and means that the initial assumption is wrong.

Hence $\sqrt{3}$ is irrational.

Note that in Section 1.2 Example 1 we used the fact that if a^2 has a factor 3, then a has a factor 3. This result was established in Example 1.

In future work you may use (without proof) that if a^2 has a factor p, p being a **prime**, then a has a factor p. For example, if p^2 is divisible by 7 then p is divisible by 7.

---

2   Complete the following proof by contradiction to show that $\sqrt{7}$ is irrational.

     Assuming that $\sqrt{7}$ is rational. Then $\sqrt{7}$ may be written in the form $\frac{a}{b}$ where a, b are integers having no common factors.

     $\therefore$ $\frac{a}{b} = \sqrt{7}$

     $\therefore$ $a = \sqrt{7}b$

     $\therefore$ $a^2 = 7b^2$, has a factor 7.

     $\therefore a$, has a factor 7 so that $a = 7k$, where k is an integer.

Answer

2   Now $a^2 = 7b^2$

     $\therefore$ $(7k)^2 = 7b^2$

     $\therefore$ $49k^2 = 7b^2$

     $\therefore$ $7k^2 = b^2$, so b^2 has a factor 7.

     $\therefore b$ has a factor 7

a and b have 7 as a common factor

This is a contradiction and means that the initial assumption is wrong.

Hence $\sqrt{7}$ is irrational.

1.4 Proof of the infinity of primes

There is no maximum prime number. Here is one way this can be proved.

Suppose we have the following list of prime numbers: $p_1, p_2, p_3, p_4, p_5, \ldots p_n$

When all of them are multiplied together and 1 is added, the resulting answer, which we can call q, may or may not be prime. If it is prime it is a prime number that was not in the original list and so is a new prime number. If it is not prime, it must be divisible by a prime number r.

Now r cannot be p_1, p_2, p_3, etc., as dividing q by any of these numbers would result in a remainder of 1. This means that r is a new prime number.

So there can be a new prime number q or if q is not prime it has a new prime for a prime factor. Hence there are an infinite number of primes.

1.5 Application of proof by contradiction to unfamiliar proofs

Some of the questions on proofs will refer to proofs you are familiar with. For example, the proof for proving $\sqrt{7}$ is irrational is much the same as the familiar proof that $\sqrt{2}$ is irrational.

You will get questions in the exam where proof by contradiction needs to be used in unfamiliar situations. Sometimes these questions are in questions that are not entirely on proof. For example, a question on trigonometry may have a part where you have to prove a particular statement. The examples below give you a good range of these types of questions.

Example

1   If a and b are real integers, then $a^2 - 4b \neq 2$.

Use proof by contradiction to prove that the above statement is true.

. . . .

Answer

1   Suppose we assume that this statement is false.

Assuming there are values for a and b such that $a^2 - 4b = 2$

Rearranging gives $\qquad a^2 = 4b + 2$

As $a^2 = 2(2b + 1)$ it means that a^2 must be even (as it has 2 as a factor). This also means a must be even as if you square an odd number you always obtain an odd number and squaring even numbers results in even numbers.

We can write $a = 2c$ for a certain integer c.

Substituting $a = 2c$ into the formula, we obtain $4c^2 = 2(2b + 1)$

Rearranging, gives $4c^2 - 4b = 2$

Dividing both sides by 2 gives $2c^2 - 2b = 1$ so $1 = 2(c^2 - b)$

Now as b and c are both integers this cannot be true as $2(c^2 - b)$ is even and 1 is odd.

A contradiction has been found, so the original assumption $a^2 - 4b = 2$ is false, so $a^2 - 4b \neq 2$ is true.

Here we assume that the initial statement in the question is false.

2 Complete the following proof by contradiction to show that

$\sin \theta + \cos \theta \le \sqrt{2}$ for all values of θ.

Assume that there is a value of θ for which $\sin \theta + \cos \theta > \sqrt{2}$

Then squaring both sides, we have: ...

Answer

2 Assume that there is a value of θ for which $\sin \theta + \cos \theta > \sqrt{2}$

Then squaring both sides, we have:

$$(\sin \theta + \cos \theta)^2 > 2$$

$$\sin^2 \theta + 2 \sin \theta \cos \theta + \cos^2 \theta > 2$$

Now $\sin^2 \theta + \cos^2 \theta = 1$

So $1 + 2 \sin \theta \cos \theta > 2$, giving $2 \sin \theta \cos \theta > 1$

Now $2 \sin \theta \cos \theta = \cos 2\theta$

So $\cos 2\theta > 1$, which is a contradiction as the cosine of any angle is ≤ 1.

BOOST

Grade ⬆⬆⬆⬆

You must be able to spot trig identities and use them in proofs such as this. Remember proofs can come up as part of any question.

3 Prove by contradiction the following proposition:

When x is real and $x \ne 0$,

$$\left| x + \frac{1}{x} \right| \ge 2$$

The first two lines of the proof are given below.

Assume that there is a real value of x such that $\left| x + \frac{1}{x} \right| < 2$

Then squaring both sides, we have: ...

The modulus of x is written as $|x|$ and this means you take the numerical value of x ignoring the signs.

Answer

3 Assume that there is a real value of x such that $\left| x + \frac{1}{x} \right| < 2$

Then squaring both sides, we have:

$$\left(x + \frac{1}{x} \right)^2 < 4$$

$$x^2 + 2 + \frac{1}{x^2} < 4$$

Multiplying both sides by x^2, we obtain

$$x^4 + 2x^2 + 1 < 4x^2$$

$$x^4 - 2x^2 + 1 < 0$$

$$(x^2 - 1)(x^2 - 1) < 0$$

$(x^2 - 1)^2 < 0$, which is impossible since the square of a real number cannot be negative.

As this is a contradiction, the assumption is incorrect, so $\left| x + \frac{1}{x} \right| \ge 2$.

Notice there is an x^2. In equations such as this we can multiply both sides by x^2 to remove the fraction.

Active Learning

There is no doubt that proofs can be tricky. If you still feel a bit unsure about them why not try looking at YouTube videos on proof by contradiction.

Try to choose ones that apply only to A-level work.

Test yourself

1. Use proof by contradiction to prove that if a^2 is even, then a must be even. [3]

2. Prove by contradiction the following proposition:

    If n is a positive integer and $3n + 2n^3$ is odd, then n is odd.

    The first two lines of the proof are given below.

    It is given that $3n + 2n^3$ is odd.
    Assume that is n is even so that n = 2k. [4]

3. Prove by contradiction the following proposition:

    If a, b are positive real numbers, then $a + b \geq 2\sqrt{ab}$.

    The first line of the proof is given below

    Assume that positive real numbers a, b exist such that $a + b < 2\sqrt{ab}$. [3]

4. Prove by contradiction the following proposition:

    If a and b are odd integers such that 4 is a factor of $a - b$, then 4 is **not** a factor of $a + b$.

    The first lines of the proof are given below.

    Assume that 4 is a factor of a + b.
    Then there exists an integer, c, such that $a + b = 4c$. [3]

5. Prove by contradiction the following proposition:

    When x is real,
    $$(5x - 3)^2 + 1 \geq (3x - 1)^2$$
    The first line of the proof is given below.

    Assume that there is a real value of x such that
    $$(5x - 3)^2 + 1 < (3x - 1)^2$$ [3]

6. Show that $\sqrt{5}$ is irrational. [6]

7. Prove by contradiction the following proposition:

    When x is real and $0 \leq x \leq \frac{\pi}{2}$, $\sin x + \cos x \geq 1$ [6]

Summary

Check you know the following facts:

Real numbers are those numbers that give a positive number when squared.

Imaginary numbers when squared give a negative number (e.g. $\sqrt{-1}$ when squared gives -1).

Rational numbers can be expressed as a fraction $\left(\text{i.e. } \frac{a}{b}\right)$ where a and b are integers with no common factors.

Irrational numbers cannot be expressed as a fraction (e.g. π, $\sqrt{2}$, $\sqrt{3}$).

Proof by contradiction

Here we assume that a statement or what we want to prove is not true and then use maths techniques to show that the consequences of this are not possible. The consequences can contradict what we have just assumed to be true or something we know to be definitely true (e.g. x^2 cannot be negative for real values of x) or both. These are called contradictions to the original assumption.

2 Algebra and functions

Introduction

You may have come across some of the content of this topic in your GCSE studies. This topic looks at a variety of algebraic techniques to simplify expressions so they can be further used in other topics such as differentiation and integration. The topic also looks at functions and their features and graphs.

This topic covers the following:

2.1 Simplifying algebraic expressions

2.2 Partial fractions

2.3 Definition of a function

2.4 Domain and range of functions

2.5 The graphical representation of functions, with input x and the outputs y

2.6 Composition of functions

2.7 Inverse functions and their graphs

2.8 The graphs of inverse functions

2.9 The modulus function

2.10 Graphs of modulus functions

2.11 Combinations of the transformations on the graph $y = f(x)$

2.12 Exponential and logarithmic functions

2.1 Simplifying algebraic expressions

Some of this section will be familiar to you as it was covered as part of your GCSE work.

When you have an algebraic fraction always look to see if any part of it can be factorised. Here are a few examples that show this.

Notice 8 can be taken out as a factor of the numerator.

The $(3x - 1)$ can be cancelled in both the numerator and denominator.

The numerator and denominator are both quadratic functions so see if they will factorise and if so, factorise them. Then cancel any common factors in the numerator and denominator.

Examples

1 Simplify $\dfrac{24x - 8}{(2x + 1)(3x - 1)}$

Answer

1 $\dfrac{24x - 8}{(2x + 1)(3x - 1)} = \dfrac{8(3x - 1)}{(2x + 1)(3x - 1)}$

$= \dfrac{8}{(2x + 1)}$

2 Simplify $\dfrac{x^2 - x - 12}{x^2 + 5x + 6}$

Answer

2 $\dfrac{x^2 - x - 12}{x^2 + 5x + 6} = \dfrac{(x + 3)(x - 4)}{(x + 3)(x + 2)} = \dfrac{x - 4}{x + 2}$

2.2 Partial fractions

Suppose we want to combine two algebraic fractions to form a single fraction. The following method can be used:

$$\frac{3}{x + 3} + \frac{2}{x - 1} \equiv \frac{3(x - 1) + 2(x + 3)}{(x + 3)(x - 1)} \equiv \frac{3x - 3 + 2x + 6}{(x + 3)(x - 1)} \equiv \frac{5x + 3}{(x + 3)(x - 1)}$$

The additional line in the equality indicates an identity, i.e. a result true for all values of x.

If, however, we want to do the reverse and write $\dfrac{5x + 3}{(x + 3)(x - 1)}$ as two separate

fractions then we are said to be expressing the single fraction in terms of partial fractions.

It is obvious what the denominators of the partial fractions will be, but we will need to call the numerators A and B until their values can be determined.

Hence $\dfrac{5x + 3}{(x + 3)(x - 1)} \equiv \dfrac{A}{x + 3} + \dfrac{B}{x - 1}$

Multiplying both sides by $(x + 3)(x - 1)$ gives

$$5x + 3 \equiv A(x - 1) + B(x + 3)$$

Values of x are chosen so that the contents of one of the brackets become zero.

By letting $x = 1$ means that the first bracket becomes zero and means that the letter A is eliminated. When $x = -3$, the second bracket becomes zero eliminating B in the process.

Let $x = 1$, so $8 = 4B$ giving $B = 2$

Let $x = -3$, so $-12 = -4A$ giving $A = 3$

Hence the given fraction expressed in partial fractions is $\dfrac{3}{x + 3} + \dfrac{2}{x - 1}$

Partial fractions where there is a repeated factor in the denominator

Suppose you have to write the fraction $\dfrac{4x + 1}{(x + 1)^2(x - 2)}$ in terms of partial fractions.

There is a repeated linear factor in the denominator (i.e. $(x + 1)^2$). The repeated linear factor means one of the denominators will be $(x + 1)^2$ and another will be $(x + 1)$. There will also be a third partial fraction with denominator $(x - 2)$.

Hence the original expression can be expressed in terms of its partial fractions like this:

$$\frac{4x + 1}{(x + 1)^2(x - 2)} \equiv \frac{A}{(x + 1)^2} + \frac{B}{x + 1} + \frac{C}{x - 2}$$

Now multiply through by the denominator of the left-hand side. This will remove the fractions.

$$4x + 1 \equiv A(x - 2) + B(x + 1)(x - 2) + C(x + 1)^2$$

Let $x = 2$, so

$$4(2) + 1 = A(2 - 2) + B(2 + 1)(2 - 2) + C(2 + 1)^2$$

$$9 = 9C$$

$$C = 1$$

Let $x = -1$, so

$$-3 = -3A, \text{ giving } A = 1$$

Let $x = 0$, so

$$1 = -2A - 2B + C$$

Now as $A = 1$ and $C = 1$, substituting these values into the above equation gives:

$$1 = -2 - 2B + 1$$

Hence,

$$B = -1$$

It is a good idea to check the partial fractions by putting a number in for x on both sides of the equation and checking to see if the right-hand side of the equation equals the left-hand side of the equation.

$$\frac{4x + 1}{(x + 1)^2(x - 2)} = \frac{1}{(x + 1)^2} - \frac{1}{x + 1} + \frac{1}{x - 2}$$

Let $x = 1$, so LHS $= -\dfrac{5}{4}$ and RHS $= \dfrac{1}{4} - \dfrac{1}{2} - 1 = -\dfrac{5}{4}$

LHS = RHS

Hence the partial fractions are $\dfrac{1}{(x + 1)^2} - \dfrac{1}{x + 1} + \dfrac{1}{x - 2}$

Examples

1 Express $\dfrac{7x^2 - 2x - 3}{x^2(x - 1)}$ in terms of partial fractions.

. . . .

Answer

1   $\dfrac{7x^2 - 2x - 3}{x^2(x - 1)} \equiv \dfrac{A}{x^2} + \dfrac{B}{x} + \dfrac{C}{x - 1}$

$7x^2 - 2x - 3 \equiv A(x - 1) + Bx(x - 1) + Cx^2$

Let $x = 1$, so $C = 2$

Let $x = 0$, so $A = 3$

Let $x = 2$, so $B = 5$

$$\frac{7x^2 - 2x - 3}{x^2(x - 1)} \equiv \frac{3}{x^2} + \frac{5}{x} + \frac{2}{x - 1}$$

If the denominator is multiplied out, the highest power of x is 3. This is the same as the required number of constants.

Alternatively, equate coefficients of x^2.

$$0 = B + C$$

so $B = -C = -1$

Do not let x equal a value that would make one of the denominators zero.

BOOST
Grade ⇧⇧⇧⇧

Always perform checks like this as it is easy to make a mistake and often the mistake would lead to further problems later on in the question.

Notice that x^2 in the denominator of the original fraction is a repeated linear factor.

Alternatively, equate coefficients of x^2.

$$7 = B + C$$

So $B = 7 - C = 5$

Partial fractions are often only one part of a question so it is important to spend the time checking them. Any value of x, other than those used already can be used for the check.

Check using $x = 3$:   LHS $= \dfrac{7(3)^2 - 2(3) - 3}{3^2(3-1)} = 3$

RHS $= \dfrac{3}{3^2} + \dfrac{5}{3} + \dfrac{2}{3-1} = 3$

Hence, partial fractions are $\dfrac{3}{x^2} + \dfrac{5}{x} + \dfrac{2}{x-1}$

---

2   The function f is defined by

$$f(x) = \frac{8 - x - x^2}{x(x-2)^2}$$

(a)   Express $f(x)$ in terms of partial fractions.

(b)   Use your results to part (a) to find the value of $f'(1)$

. . . .

Answer

The $(x-2)^2$ is a repeated linear factor.

2   (a) $\dfrac{8 - x - x^2}{x(x-2)^2} \equiv \dfrac{A}{x} + \dfrac{B}{(x-2)^2} + \dfrac{C}{x-2}$

$x = 2$ means two of the brackets will equal zero.

$8 - x - x^2 \equiv A(x-2)^2 + Bx + Cx(x-2)$ Multiply both sides by $x(x-2)^2$

Let $x = 2$, $2 = 2B$, so $B = 1$

Let $x = 0$, $8 = 4A$, so $A = 2$

Substituting a value such as $x = 1$ into both sides of the equation acts as a check on the values of A, B and C. If the left-hand side equals the right-hand side, then there is a good chance the values are correct.

Let $x = 3$, $-4 = A + 3B + 3C$, so $C = -3$ Substitute the values already obtained for A and B to find C or equate coefficients of x^2.

Hence $\dfrac{8 - x - x^2}{x(x-2)^2} \equiv \dfrac{2}{x} + \dfrac{1}{(x-2)^2} - \dfrac{3}{x-2}$

Check using $x = 1$:   LHS $= 6$ RHS $= 6$

(b) $f(x) = \dfrac{2}{x} + \dfrac{1}{(x-2)^2} - \dfrac{3}{x-2}$ Express the fractions in index form to enable differentiation.

The Chain rule is used to differentiate the last two terms.

$f(x) = 2x^{-1} + (x-2)^{-2} - 3(x-2)^{-1}$

$f'(x) = -2x^{-2} - 2(x-2)^{-3} + 3(x-2)^{-2}$

Change back to algebraic fractions from index form to make it easy for numbers to be substituted in for x.

$f'(x) = -\dfrac{2}{x^2} - \dfrac{2}{(x-2)^3} + \dfrac{3}{(x-2)^2}$

$f'(1) = -\dfrac{2}{1^2} - \dfrac{2}{(1-2)^3} + \dfrac{3}{(1-2)^2} = -2 + 2 + 3 = 3$

---

3   Express $\dfrac{5x^2 - 8x - 1}{(x-1)^2(x-2)}$ in terms of partial fractions.

. . . .

Answer

3   Let $\dfrac{5x^2 - 8x - 1}{(x-1)^2(x-2)} \equiv \dfrac{A}{(x-1)^2} + \dfrac{B}{(x-1)} + \dfrac{C}{x-2}$

$5x^2 - 8x - 1 \equiv A(x-2) + B(x-1)(x-2) + C(x-1)^2$

Let $x = 2$,   so $C = 3$

Let $x = 1$,   so $A = 4$

Alternatively, equate coefficients of x^2.

$5 = B + C$

So $B = 5 - C = 2$

Let $x = 0$, giving $-1 = -2A + 2B + C$ giving $B = 2$

Hence $\dfrac{5x^2 - 8x - 1}{(x-1)^2(x-2)} = \dfrac{4}{(x-1)^2} + \dfrac{2}{x-1} + \dfrac{3}{x-2}$

Checking by letting $x = 3$:

$$\text{LHS} = \dfrac{5(3)^2 - 8(3) - 1}{(3-1)^2(3-2)} = 5$$

$$\text{RHS} = \dfrac{4}{(3-1)^2} + \dfrac{2}{3-1} + \dfrac{3}{3-2} = 1 + 1 + 3 = 5$$

Hence $\qquad$ LHS = RHS

Partial fractions are $\dfrac{4}{(x-1)^2} + \dfrac{2}{x-1} + \dfrac{3}{x-2}$

2.3 Definition of a function

Look at the diagram below where each element of a given set {1, 3, 5} is mapped to one element in another set, {3, 9, 15}.

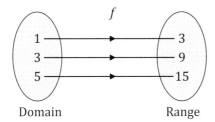

A **function** is a relation between a set of inputs and a set of outputs such that each input is related to exactly one output.

The given set (i.e. {1, 3, 5} here) is called the **domain** of the function and the set to which the domain is mapped (i.e. {3, 9, 15} here) is called the **range** of the function.

The above diagram could be represented by the following mathematical rule:

$f(x) = 3x$ with the domain $x \in \{1, 3, 5\}$ and the range $f(x) \in \{3, 9, 15\}$

A function maps each input to only one output

This is best explained by considering the following example.

$y = x^2$ is a function because whatever value x takes there is only one possible y value. For example, if $x = 3$ then $y = 9$.

However, if you consider the mapping $y = \pm\sqrt{x}$, and substitute $x = 4$, then there are two possible values of y, (i.e. 2 or -2), so $y = \pm\sqrt{x}$ is not a function. Hence $y = x^2$ is a function, whereas $y = \pm\sqrt{x}$ is not.

2.4 Domain and range of functions

All functions have a domain and a range.

> The domain is the set of input values that can be entered into the function and the range is the set of output values.

Interval notation

Domains and ranges can be expressed as intervals, using the following notation:

- The use of parentheses: (a, b) means the open interval $a < x < b$ (**not including** the endpoints).
- The use of square brackets: $[a, b]$ means the closed interval $a \leq x \leq b$ (**including** the endpoints).

So, $(a, b]$ means $a < x \leq b$ and $[a, b)$ means $a \leq x < b$.

The smallest value appears first and then the largest value, i.e. $(-1, 4)$, not $(4, -1)$.

For example:

$(-1, 4)$ means all numbers between -1 and 4, not including -1 and 4.

$[-1, 4]$ means all numbers between -1 and 4, including -1 and 4.

$[-1, 4)$ means all numbers between -1 and 4, including -1 but not 4.

$(-\infty, 4]$ means all numbers less than or equal to 4.

$(-1, \infty)$ means all numbers greater than -1.

2.5 The graphical representation of functions, with input x and the outputs y

To understand functions you need to be able to sketch graphs. Graph sketching was covered in Unit 1.

If you are asked to find the largest possible domain for a function then you should sketch a graph of the function to see what input values are allowable.

Take the function $f(x) = 2x^2 + 3$ as an example. You can sketch this curve by considering the transformations to the graph $y = x^2$, to produce $f(x) = 2x^2 + 3$ (i.e. a stretch parallel to the y-axis with scale factor 2 and a translation of three units up). The following graph is obtained:

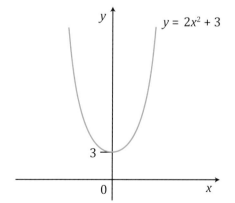

From the graph it can be seen all x-values are allowable as inputs, so the largest domain is from $-\infty$ to ∞ which can be written as $D(f) = (-\infty, \infty)$. The outputs (i.e. the y-values) are all values greater than or equal to 3, which is the range and can be written as $R(f) = [3, \infty)$.

Example

1 The function f has domain $(-\infty, -1]$ and is defined by

$$f(x) = 3x^2 - 2$$

Notice the way the domain is written. The brackets show that the domain is all numbers less than or equal to −1.

You will often see the domain written in this form: $D(f) = (-\infty, -1]$.

To find the range, you need to find the least and greatest values of $f(x)$, for values of x that are within the domain of f.

If you sketch the curve $y = 3x^2 - 2$ and then mark the domain, you can work out the range in the following way.

> Round brackets are **always** used for ∞ or $-\infty$, since, by definition, we can never get there. Instead, we should consider approaching ∞ or $-\infty$.

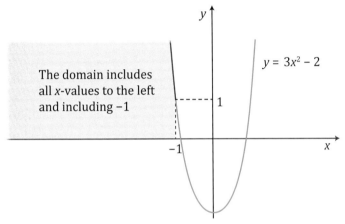

The domain includes all x-values to the left and including −1

$y = 3x^2 - 2$

To find the y-values on the graph, the x-coordinates are substituted into the function (or the equation of the curve) like this:

$$f(-1) = 3(-1)^2 - 2 = 3 - 2 = 1$$

We can't substitute ∞ into the function, as it is not a number. Instead, we need to think about the behaviour of the function as x gets closer to ∞.

As x becomes large and negative, x^2 becomes large and positive. Multiplying by 3 and subtracting 2 make little difference as the value of x^2 gets larger. Therefore we say that as $x \to \infty, f(x) \to \infty$.

You can now see that only the part of the curve shown in red is allowed owing to the restricted domain. The range of the function (i.e. the allowable $f(x)$ or y-values) is the set of numbers greater than or equal to 1, and this can be written in shorthand as $R(f) = [1, \infty)$.

2.6 Composition of functions

Composition of functions involves applying two or more functions in succession. To understand this, consider the following example.

$$\text{If } f(x) = x^2 \text{ and } g(x) = x - 2$$

$fg(x)$ results from replacing x by the expression for $g(x)$ in $f(x)$.

> Here, g means 'subtract 2 from it', and f means 'square it'. So fg means 'subtract 2 from it and then square it', i.e. $(x - 2)^2$, and gf means 'square it and then subtract 2 from it', i.e. $x^2 - 2$.

> The composite function $fg(x)$ means $f(g(x))$ and is the result of performing the function g first and then f.

That is: $$fg(x) = f(g(x)) = f(x - 2) = (x - 2)^2$$

The domain and range of composite functions

Provided that a composite function exists, the domain of the composite function $fg(x)$ is the set of x values in the domain of g for which $g(x)$ is in the domain of f.

To find the domain of a composite function you need to find the domain of the first function (i.e. the function nearest the x) and then the composed function.

For example, when finding the domain of the composite function $gf(x)$ you need to:

- First find the domain for $f(x)$
- Check that $g(x)$ is given the correct domain.

Suppose you have the functions $f(x) = \sqrt{x}$ and $g(x) = x^2$, and are asked to find the domain of the composite function $gf(x)$.

The domain of $f(x) = \sqrt{x}$ is $D(f) = (0, \infty)$ (i.e. all non-negative real numbers).

The composite function $gf(x) = (\sqrt{x})^2 = x$

Now this composite function (i.e. x) would normally have the set of all real numbers as its domain, but in this case we have to consider the domain of $f(x)$ which will restrict the domain of the composite function. As only non-negative numbers can be put into $f(x)$ this will mean only these numbers can be used as the domain for the composite function $gf(x)$.

The domain of the composite function $gf(x)$ is $D(gf) = (0, \infty)$

Example

1   If $f(x) = x^2 - 3$ and $g(x) = \sqrt{x - 2}$, find the domain of $fg(x)$.

· ·

Answer

1   First consider the domain for $g(x)$ and as $g(x) = \sqrt{x - 2}$ we cannot find the square root of a negative number, so this means that the value of x has to be 2 or more.

Now $fg(x) = \left(\sqrt{x - 2}\right)^2 - 3 = x - 2 - 3 = x - 5$ and as $x - 5$ has the domain of all real numbers (i.e. $(-\infty, \infty)$), but as the domain is restricted by $g(x)$ the domain is $D(fg) = [2, \infty)$.

$f(x) = x^2 - 3$ would have the domain of all real numbers (i.e. $(-\infty, \infty)$).

In general, the domain of a composite function is either the same as the domain of the first function or else it lies inside it. So for a composite function $fg(x)$, the domain would be either the domain of g or part of the domain of g.

Rule for a composite function to exist

Not all composite functions exist. For the composite function fg to exist, check that the range of g (from its graph) is a subset of, or an equal set to, the domain of f.

Examples

1   If $f(x) = x^2$ and $g(x) = x - 6$ find

(a) $fg(x)$

(b) $gf(x)$

Answer

   (a) $fg(x) = (x - 6)^2$

   (b) $gf(x) = x^2 - 6$

---

2 The functions f and g have domains $[-3, \infty)$ and $(-\infty, \infty)$ respectively and are defined by

$$f(x) = \sqrt{x + 4}$$

$$g(x) = 2x^2 - 3$$

   (a) Write down the range of f and the range of g.

   (b) Find an expression for $gf(x)$. Simplify your answer.

   (c) Solve the equation $fg(x) = 17$.

Answer

2   (a) Now $f(x) = \sqrt{x + 4}$

      When $x = -3$, $f(-3) = \sqrt{-3 + 4} = 1$

      As $x \to \infty$ the value of $f(x)$ gets larger, so $f(x) \to \infty$

      Hence $R(f) = [1, \infty)$

      Now $g(x) = 2x^2 - 3$

      When $x = 0$, $g(0) = 2(0)^2 - 3 = -3$

      As $x \to \infty$ the value of $g(x)$ gets larger, so $g(x) \to \infty$

      $R(g) = [-3, \infty)$

   (b) $gf(x) = 2\left(\sqrt{x + 4}\right)^2 - 3$

          $= 2(x + 4) - 3$

          $= 2x + 5$

   (c) $fg(x) = \sqrt{(2x^2 - 3 + 4)}$

         $= \sqrt{(2x^2 + 1)}$

      Now $fg(x) = 17$.

$$17 = \sqrt{(2x^2 + 1)}$$
$$289 = 2x^2 + 1$$
$$288 = 2x^2$$
$$144 = x^2$$
$$x = \pm 12$$

To find the least value of $f(x)$, the contents of the square root must be as small as possible. By inspection you can see that this occurs when $x = -3$. Note that the least value of $\sqrt{x + 4}$ occurs when $x = -4$, but this is not in the domain of f, which states that x must be greater than or equal to -3.

$g(x)$ has its least value when $x = 0$. You can obtain this value by thinking about the minimum point if the function were plotted. Always check that the value of x lies within the domain (i.e. $(-\infty, \infty)$ in this case).

Instead of using interval notation, it would also be acceptable to describe the range using words or inequality symbols, e.g. the range of g is all numbers greater than or equal to -3, or $x \geq -3$, or $g(x) \geq -3$.

Square both sides to remove the square root.

Remember to include $\pm$ when square rooting, and then check whether one or both values are in the domain of fg, i.e. the domain of g. In this case, both values are needed.

2.7 Inverse functions and their graphs

To check that a function has an inverse function we check to see if the function is one-to-one. If you sketch the graph of the function and if a horizontal line cuts the graph in more than one place it is not one-to-one and doesn't have an inverse function.

A function f produces a single output from an input. The inverse function of f reverses the process: it obtains the input of f from a given output. The inverse of the function f is written as f^{-1}. For the inverse f^{-1} to exist the function f must be one-to-one; otherwise there would be two inputs of f corresponding to a given output of f and it would be impossible to determine which input value of f (which is an output for f^{-1}) would be appropriate.

The domain of f (i.e. set of inputs) is identical to the range (i.e. set of outputs) of the inverse function f^{-1}.

To find the inverse of a function, follow these steps:

1   Let the function equal y.
2   Rearrange the resulting equation so that x is the subject of the equation.
3   Replace x with $f^{-1}(x)$ and replace y with x.

These steps are shown in the following example:

The function f has domain $[0, \infty)$ and is defined by:

$$f(x) = 5x^2 + 3 \text{ and you are asked to find } f^{-1}(x).$$

Rearrange the resulting equation so that x is the subject of the equation. When square rooting there would normally be a ± placed before the root. However, here only the positive value is allowed because the domain of f is $[0, \infty)$.

Step 1 Let $y = 5x^2 + 3$ Let the function equal y.

Step 2 $x = \sqrt{\dfrac{y - 3}{5}}$

Step 3 $f^{-1}(x) = \sqrt{\dfrac{x - 3}{5}}$ Replace x with $f^{-1}(x)$ and replace y with x.

Examples

1   The function f has domain $(-\infty, -1]$ and is defined by

$$f(x) = 2x^2 - 1$$

(a) Write down the range of f.

(b) Find $f^{-1}(x)$.

· ·

Answer

If the graph of $y = 2x^2 - 1$ is drawn, the minimum point occurs at $(0, -1)$. $x = 0$ is not allowable as it lies outside the domain for the function (i.e. $D(f) = (-\infty, -1])$. The largest allowable x value is therefore -1 and the allowable part of the curve is all points including and to the left of this value (i.e. from -1 to $-\infty$). The corresponding y-values (i.e. the range) are obtained by substituting these two values for the domain into the equation for the curve.

1   (a) $f(x) = 2x^2 - 1$

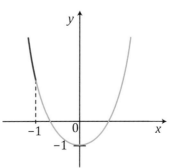

$f(-1) = 2(-1)^2 - 1 = 2 - 1 = 1$

As $x \to -\infty$, $f(x) \to \infty$

Hence $R(f) = [1, \infty)$

(b) Let $y = 2x^2 - 1$

$$x^2 = \frac{y + 1}{2}$$

$$x = -\sqrt{\frac{y + 1}{2}}$$

$$f^{-1}(x) = -\sqrt{\frac{x + 1}{2}}$$

> The negative square root needs to be used here because the domain of f or range of f^{-1} is $(-\infty, -1]$.

---

2 The function f has domain $(-\infty, -1]$ and is defined by

$$f(x) = 4x^2 - 3$$

(a) Write down the range of f.

(b) Find an expression for $f^{-1}(x)$ and write down the range and domain of f^{-1}.

(c) (i)   Evaluate $f^{-1}(6)$.

  (ii) By carrying out an appropriate calculation involving f, verify that your answer to part (i) is correct.

· ·

Answer

2 (a) As $x \to -\infty$, $f(x) \to \infty$

  When $x = -1, f(-1) = 4(-1)^2 - 3 = 1$

  $R(f) = [1, \infty)$

> The least value $f(x)$ can take is when $x = -1$. Note that this value of x lies in the domain of the function.

> Draw or imagine the graph for $y = x^2 - 3$. It will have a minimum point at $(0, -3)$. However, $x = 0$ does not lie in the domain.

(b) Let $y = 4x^2 - 3$

$$\frac{y + 3}{4} = x^2$$

$$x = \pm\frac{1}{2}\sqrt{(y + 3)}$$

> Rearrange for x^2 and then square root and then determine whether to use the + or − sign.

According to the domain of f (i.e. range of f^{-1}), x could only take the negative value.

$$f^{-1}(x) = -\frac{1}{2}\sqrt{x + 3}$$

$$R(f^{-1}) = (-\infty, -1]$$

$$D(f^{-1}) = [1, \infty)$$

> The range of f^{-1} is the same as the domain of $f(x)$.

(c) (i) $f^{-1}(6) = -\frac{1}{2}\sqrt{6 + 3} = -\frac{3}{2}$

> $x = 6$ is substituted into $f^{-1}(x)$.

  (ii) $f\left(-\frac{3}{2}\right) = 4\left(-\frac{3}{2}\right)^2 - 3 = 4\left(\frac{9}{4}\right) - 3 = 6$

> Note the function turns the value $\left(-\frac{3}{2}\right)$ into 6 and the inverse function does the reverse, changing 6 into $\left(-\frac{3}{2}\right)$. This can be used to check that the inverse is correct.

3   The function f has domain $(-\infty, -1]$ and is defined by

$$f(x) = 6x^2 - 2$$

(a)   Write down the range of f.

(b)   Find an expression for $f^{-1}(x)$ and write down the range and domain of f^{-1}.

(c)   (i)   Evaluate $f^{-1}(3)$.

(ii)   By carrying out an appropriate calculation involving f, verify that your answer to part (i) is correct.

. . . .

Consider the values of x in the domain of f that will give the greatest and least values for $f(x)$.

Answer

3   (a)   As $x \rightarrow -\infty$, $f(x) \rightarrow \infty$

$f(-1) = 6(-1)^2 - 2 = 4$

Hence $R(f) = [4, \infty)$

The least value in the range is 4. It is possible to have the exact value so a square bracket is used.

(b)   Let $y = 6x^2 - 2$

$$x^2 = \frac{y + 2}{6}$$

$$x = -\sqrt{\frac{y + 2}{6}}$$

$$f^{-1}(x) = -\sqrt{\frac{x + 2}{6}}$$

The negative square root is taken because the domain of f (and therefore the range of f^{-1}) is $(-\infty, -1]$.

Range of f^{-1} is the same as the domain of f.

Hence $R(f^{-1}) = (-\infty, -1]$

Domain of f^{-1} is the same as the range of f.

Hence $D(f^{-1}) = [4, \infty)$

(c)   (i) $f^{-1}(x) = -\sqrt{\dfrac{x + 2}{6}}$

$$f^{-1}(3) = -\sqrt{\frac{3 + 2}{6}} = -\sqrt{\frac{5}{6}}$$

(ii)   If $x = -\sqrt{\dfrac{5}{6}}$ is put back into the original function the answer should be 3.

Now $f(x) = 6x^2 - 2 = 6\left(-\sqrt{\dfrac{5}{6}}\right)^2 - 2 = 6 \times \dfrac{5}{6} - 2 = 3$

Hence the answer to part (i) is correct.

4   The function f has domain $x \leq 0$ and is defined by $f(x) = 5x^2 + 4$

(a)   Find an expression for $f^{-1}(x)$.

(b)   Write down the domain and range of f^{-1}.

Answer

Rearrange the equation so that x becomes the subject.

4 (a) Let $y = 5x^2 + 4$

$$\frac{y - 4}{5} = x^2$$

Hence $x = \pm\sqrt{\dfrac{y - 4}{5}}$

When square rooting you must remember to include $\pm$.

However, as $x \le 0$, $x = -\sqrt{\dfrac{y - 4}{5}}$

So $f^{-1}(x) = -\sqrt{\dfrac{x - 4}{5}}$

Check with the domain of f to see which of the values is required. Note that the domain of f here only allows a value of x which is 0 or negative. Hence the positive sign cannot be used.

(b) Domain of $f^{-1} = D(f^{-1}) = R(f) = [4, \infty)$ or $x \ge 4$

Range of $f^{-1} = R(f^{-1}) = D(f) = (-\infty, 0]$ or $f^{-1}(x) \le 0$

The domain of f^{-1} is the same as the range of f, which is $[4, \infty)$.

2.8 The graphs of inverse functions

To obtain the graph of an inverse function, reflect the original graph in the line $y = x$. When drawing the graph the same scale must be used on both axes otherwise the graph would become distorted.

For example, the graph below shows the original function $y = 2x + 5$ and the inverse function $y = \dfrac{x - 5}{2}$ along with the line $y = x$. Notice how the function and its inverse are reflections in the line $y = x$.

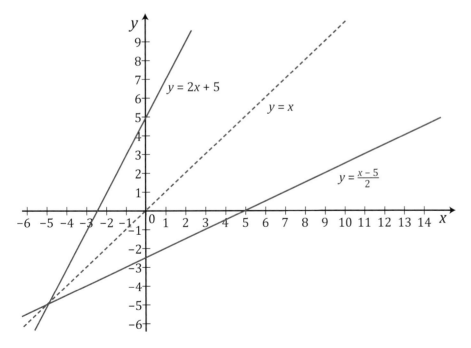

Thus to obtain the graph of $y = f^{-1}(x)$ reflect the graph of $y = f(x)$ in the line $y = x$.

2.9 The modulus function

The modulus of x is written $|x|$ and means the numerical value of x (ignoring the sign).

So whether x is positive or negative, $|x|$ is always positive (or zero).

So $|5| = 5$ and $|-5| = 5$.

The graph of $y = |x|$ is shown here.

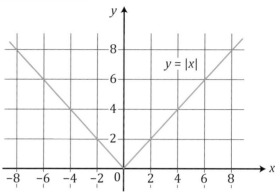

For example

$|x| = 3$ means $x = \pm 3$

$|x - 1| = 2$ means $x - 1 = \pm 2$

So $x = 2 + 1 = 3$ or $x = -2 + 1 = -1$

Hence $x = 3$ or -1

> Removal of the modulus sign in both of these examples means that a $\pm$ sign must be included as shown.

Examples

1   Solve the following

$$3|x - 5| = 9$$

....

Answer

1
$$3|x - 5| = 9$$
$$|x - 5| = 3$$
$$x - 5 = \pm 3$$
$$x = 3 + 5 = 8 \quad \text{or } x = -3 + 5 = 2$$

Hence $x = 2$ or 8

> Treat this just like an ordinary equation and make $|x - 5|$ the subject of the equation.

> Removal of the modulus sign on the left means that a $\pm$ sign must be inserted on the right.

---

2   Solve $3|x + 1| - 4 = 8$

....

Answer

2
$$3|x + 1| = 12$$
$$|x + 1| = 4$$
$$x + 1 = \pm 4$$
$$x + 1 = 4, \ \text{so } x = 3$$

Or
$$x + 1 = -4, \ \text{so } x = -5$$

Hence $x = 3$ or -5

3 Solve the following $\dfrac{3|x| - 1}{|x| + 1} = 2$

Answer

3
$$\dfrac{3|x| - 1}{|x| + 1} = 2$$

$$3|x| - 1 = 2|x| + 2$$

$$|x| = 3$$

$$x = \pm 3$$

Multiply both sides by the denominator, $|x| + 1$.

4 Solve the following

(a) $2|x + 1| - 3 = 7$

(b) $|5x - 8| \geq 3$

Answer

4 (a) $2|x + 1| - 3 = 7$

$$2|x + 1| = 10$$

$$|x + 1| = 5$$

$$x + 1 = \pm 5$$

$$x + 1 = 5 \ \text{ or } \ x + 1 = -5$$

Hence $x = 4$ or $x = -6$

(b) $|5x - 8| \geq 3$

If you have an inequality $|x| \geq a$, the solution always starts by splitting the inequality into two pieces: $x \leq -a$ or $x \geq a$. Hence we can write the inequality here as the following two pieces:

$$5x - 8 \geq 3 \ \text{ or } \ 5x - 8 \leq -3$$

$$5x \geq 11 \ \text{ or } \ 5x \leq 5$$

$$x \geq \dfrac{11}{5} \ \text{ or } \ x \leq 1$$

(b) *Alternative method*

$$|5x - 8| \geq 3$$

Squaring both sides gives:

$$(5x - 8)^2 \geq 9$$

$$25x^2 - 80x + 64 \geq 9$$

$$25x^2 - 80x + 55 \geq 0$$

$$5x^2 - 16x + 11 \geq 0$$

$$(5x - 11)(x - 1) \geq 0$$

The critical values of x are $x = \dfrac{11}{5}$ and $x = 1$

Required range is $x \leq 1$ and $x \geq \dfrac{11}{5}$

Divide through by 5 which will make the resulting equation easier to factorise.

When the contents inside the modulus sign are squared the modulus sign can be removed.

If a graph of
$$y = (5x - 11)(x - 1)$$
is plotted, the curve will be ∪-shaped, cutting the x-axis at $x = \dfrac{11}{5}$ and $x = 1$.
The section of the graph needed will be above the x-axis and the required range will be less than or equal to the lower root (1), **or** greater than, or equal to the higher root $\left(\dfrac{11}{5}\right)$.

31

5   Solve the following

$$7|x| - 2 = 8 - 3|x|$$

. . . .

Answer

5        $7|x| - 2 = 8 - 3|x|$

$$10|x| = 10$$

$$1|x| = 1$$

$$x = \pm 1$$

---

6   Solve the following

$$|5x - 2| > 8$$

. . . .

Answer

6            $|5x - 2| > 8$

$$5x - 2 > 8 \text{ or } 5x - 2 < -8$$

$$5x > 10 \text{ or } 5x < -6$$

$$x > 2 \text{ or } x < -\frac{6}{5}$$

6 *Alternative method*

The following alternative method can also be used.

$$|5x - 2| > 8$$

> The modulus sign can be removed by squaring both sides.

Squaring both sides gives:

$$(5x - 2)^2 > 64$$

$$25x^2 - 20x + 4 > 64$$

$$25x^2 - 20x - 60 > 0$$

> Here the quadratic is divided by 5 to make the result easier to factorise.

$$5x^2 - 4x - 12 > 0$$

$$(5x + 6)(x - 2) > 0$$

> Find the critical values by putting $(5x + 6)(x - 2) = 0$ and then solving.

The critical values are $x = -\dfrac{6}{5}$ and $x = 2$

If the graph of $y = (5x + 6)(x - 2)$ were plotted the curve would intersect the x-axis at $x = -\frac{6}{5}$ and $x = 2$. The graph would be ∪-shaped and the part of the graph above the x-axis would represent $(5x + 6)(x - 2) > 0$.

> A mark will be lost if 'and' is written instead of 'or' here. The reasoning for this is that x cannot be both less than $-\frac{6}{5}$ and greater than 2.

Hence $x < -\dfrac{6}{5}$ **or** $x > 2$

---

7   Solve the following

(a) $|9x - 7| \leq 3$

(b) $\sqrt{5|x| + 1} = 3$

Answer

7   (a) $|9x - 7| \leq 3$

$9x - 7 \leq 3$ and $9x - 7 \geq -3$

$9x \leq 10$ and $9x \geq 4$

$x \leq \dfrac{10}{9}$ and $x \geq \dfrac{4}{9}$

$\dfrac{4}{9} \leq x \leq \dfrac{10}{9}$

Note that the word 'and' should be included.

7   (a)   *Alternative method*

The following alternative method can also be used.

$$|9x - 7| \leq 3$$

Squaring both sides gives:

$$81x^2 - 126x + 49 \leq 9$$

$$81x^2 - 126x + 40 \leq 0$$

$$(9x - 10)(9x - 4) \leq 0$$

At the critical points $x = \dfrac{10}{9}$ or $x = \dfrac{4}{9}$

Hence $\dfrac{4}{9} \leq x \leq \dfrac{10}{9}$

The modulus sign can be removed by squaring both sides.

(b) $\sqrt{5|x| + 1} = 3$

Squaring both sides gives:

$$5|x| + 1 = 9$$

$$5|x| = 8$$

$$|x| = \dfrac{8}{5}$$

$$x = \pm\dfrac{8}{5}$$

If the curve
$$y = (9x - 10)(9x - 4)$$
were plotted it would be U-shaped and would cut the x-axis at these two points. As we want the values of x for which $y \leq 0$, this is the part of the curve below the x-axis, i.e. the values of x that are between the two roots.

2.10 Graphs of modulus functions

The graph of $y = |f(x)|$ is the graph of $y = f(x)$ with any parts of the graph below the x-axis reflected in the x-axis.

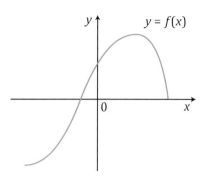

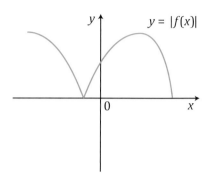

Active Learning

You can use Google to display graphs of modulus functions. Try it and see.

To draw the graph of $y = |x - 3|$ you can draw the graph of $y = x - 3$ by first finding the coordinates of intersection with each axis and then reflecting those parts which are below the x-axis in the x-axis.

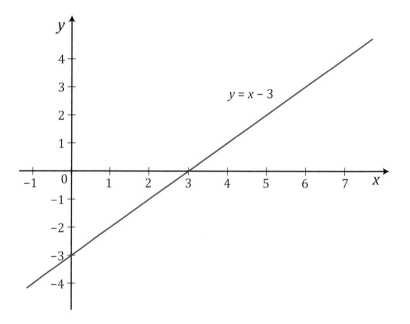

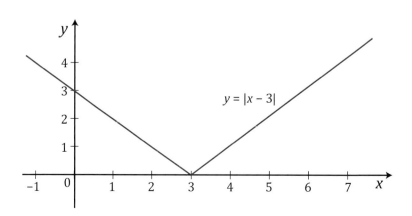

Examples

1 Given that $f(x) = 3x + 4$

    (a) Find $f^{-1}(x)$

    (b) Sketch the graph of $f(x)$ and $f^{-1}(x)$ on the same set of axes.

. . . .

Answer

1   **(a)**   Let $y = 3x + 4$

       Rearranging for x we obtain $x = \dfrac{y - 4}{3}$

       Hence, $f^{-1}(x) = \dfrac{x - 4}{3}$

    (b)

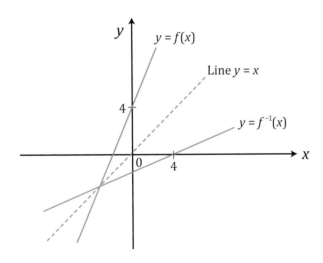

> Draw the line for the original function $y = f(x)$ and then reflect it in the line $y = x$.

---

2 The diagram shows a sketch of the graph of $y = a|x + b|$, where a and b are constants. The graph meets the x-axis at the point $(4, 0)$ and the y-axis at the point $(0, -8)$.

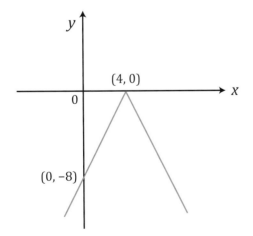

Find the value of a and the value of b.

Answer

2   As $y = a|x + b|$ and all the values of y are zero or negative, we can see that a must be negative.

$$y = a|x + b|$$

$$\frac{y}{a} = \pm(x + b)$$

When $x = 4, y = 0,$ so $0 = \pm(4 + b)$ giving $b = -4$

Hence we can write $\frac{y}{a} = \pm(x - 4)$

When $x = 0, \ y = -8,$ so $\frac{-8}{a} = \pm(0 - 4)$ giving $-8 = \pm 4a$

so $a = 2$ or -2 and, since a must be negative, $a = -2$.

Hence $a = -2$ and $b = -4$.

2.11 Combinations of the transformations on the graph $y = f(x)$

You came across the transformations of the graph $y = f(x)$ in AS Unit 1. If you are unsure about transformations of curves, you should look back at your notes.

For A2 Unit 3 you have to apply several transformations in succession.

Here is a summary of single transformations which you will need to revise.

Transformations of the graph of $y = f(x)$

A graph of $y = f(x)$ can be transformed into a new function using the rules shown in this table.

Original function	New function	Transformation
$y = f(x)$	$y = f(x) + a$	Translation of a units parallel to the y-axis, i.e. translation of $\begin{pmatrix} 0 \\ a \end{pmatrix}$
$y = f(x)$	$y = f(x + a)$	Translation of a units to the left, parallel to the x-axis, i.e. translation of $\begin{pmatrix} -a \\ 0 \end{pmatrix}$
$y = f(x)$	$y = f(x - a)$	Translation of a units to the right, parallel to the x-axis, i.e. translation of $\begin{pmatrix} a \\ 0 \end{pmatrix}$
$y = f(x)$	$y = -f(x)$	A reflection in the x-axis
$y = f(x)$	$y = af(x)$	One-way stretch with scale factor a parallel to the y-axis
$y = f(x)$	$y = f(ax)$	One-way stretch with scale factor $\frac{1}{a}$ parallel to the x-axis

Original function	New function	Transformation
$y = f(x)$	$y = f(x) + a$	
$y = f(x)$	$y = f(x + a)$	
$y = f(x)$	$y = f(x - a)$	
$y = f(x)$	$y = -f(x)$	
$y = f(x)$	$y = af(x)$ E.g. $y = 2f(x)$	
$y = f(x)$	$y = f(ax)$ E.g. $y = f(2x)$	

Examples

1 The diagram shows a sketch of the graph of $y = f(x)$. The graph passes through the points $(-2, 0)$ and $(4, 0)$ and has a maximum point at $(1, 3)$.

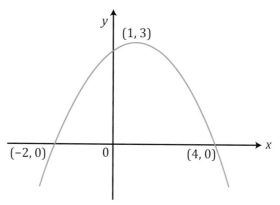

Sketch the graph of $y = -3f(x + 2)$, indicating the coordinates of the stationary point and the coordinates of the points of intersection of the graph with the x-axis.

Answer

1 There are three separate transformations:

Transformation 1:

$y = f(x)$ to $y = f(x + 2)$ represents a translation of -2 units parallel to the x-axis.

Transformation 2:

$y = f(x)$ to $y = 3f(x)$ represents a one-way stretch with scale factor 3 parallel to the y-axis.

Transformation 3:

$y = f(x)$ to $y = -f(x)$ represents a reflection in the x-axis.

Hence $y = -3f(x + 2)$ is a combination of all three transformations.

Transformation 1 means the whole graph will be shifted to the left by two units.

Transformation 2 will stretch the y-coordinates by 3 (i.e. they will be multiplied by 3), leaving the x-coordinates unchanged.

Transformation 3 will reflect the whole graph in the x-axis.

These three transformations will produce the following curve.

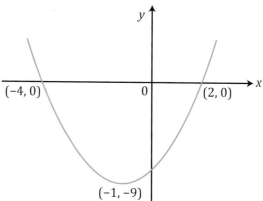

2 The function f is defined by $f(x) = |x|$.

    (a) Sketch the graph of $y = f(x)$.

    (b) On a separate set of axes, sketch the graph of $y = f(x - 5) + 3$. Mark on your sketch, the coordinates of the point on the graph where the y-coordinate is least and also the coordinates of the point where the graph crosses the y-axis.

Answer

2 (a)

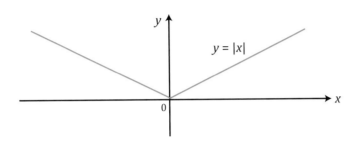

    (b) The graph of $y = f(x - 5) + 3$ can be obtained from the graph in part (a) by applying a translation of 5 units parallel to the x-axis and then a translation of 3 units parallel to the y-axis.

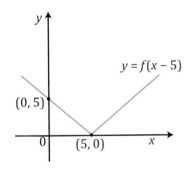 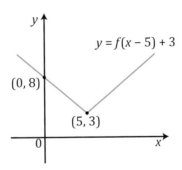

---

3 (a) Solve the inequality $|3x + 1| \leq 5$.

    (b) The function f is defined by $f(x) = |x|$.

       (i) Sketch the graph of $y = f(x)$.

       (ii) On a separate set of axes, sketch the graph of $y = f(x - 3) + 2$. On your sketch, indicate the coordinates of the point on the graph where the value of the y-coordinate is least and the coordinates of the point where the graph crosses the y-axis.

Answer

3 (a) $|3x + 1| \leq 5$

          $3x + 1 \leq 5$ and $3x + 1 \geq -5$

             $3x \leq 4$ and $3x \geq -6$

        $x \leq \dfrac{4}{3}$ and $x \geq -2$, i.e. $-2 \leq x \leq \dfrac{4}{3}$

(a) *Alternative method*

The following alternative method can also be used.

$$|3x + 1| \leq 5$$

$$(3x + 1)^2 \leq 25$$

Squaring both sides means that the modulus sign can be removed.

$$9x^2 + 6x + 1 \leq 25$$

$$9x^2 + 6x - 24 \leq 0$$

$$3x^2 + 2x - 8 \leq 0$$

Simplify the quadratic equation by dividing both sides by 3.

$$(3x - 4)(x + 2) \leq 0$$

If a graph of
$$y = (3x - 4)(x + 2)$$
were plotted it would be U-shaped and intersect the x-axis at the critical values. The required values of x would be all those for which the curve lies on or below the x-axis, i.e. the values of x between the two roots.

Critical values are $x = \dfrac{4}{3}$ and $x = -2$

Hence $-2 \leq x \leq \dfrac{4}{3}$

(b) (i)

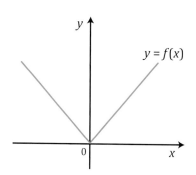

(ii)

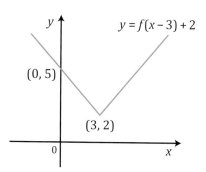

4

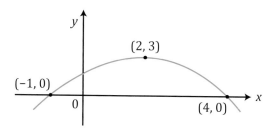

The diagram shows a sketch of the graph of $y = f(x)$. The graph has its highest point at $(2, 3)$ and intersects the x-axis at the points $(-1, 0)$ and $(4, 0)$. Sketch the graph of $y = 3f(x - 2)$, indicating the coordinates of three points on the graph.

Answer

4   $y = f(x)$ to $y = 3f(x - 2)$ represents two transformations. A translation to the right by 2 units and a one-way stretch of scale factor 3, parallel to the y-axis.

> The first transformation will increase all the x-coordinates by 2. The second transformation will result in the y-coordinates being multiplied by 3.

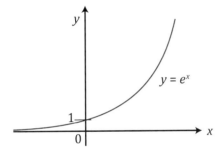

2.12 **Exponential and logarithmic functions**

The function e^x and its graph

The function $y = e^x$ has a domain of all real numbers and is a one-to-one function, meaning that one value of y corresponds to only one value of x. Hence, e^x has an inverse function. The graph of $y = e^x$ is shown below.

Notice that:

- The graph of $y = e^x$ cuts the y-axis at $y = 1$ (because when $x = 0$, $y = e^0 = 1$)
- For large negative values of x, y approaches zero from above, i.e. the x-axis is an asymptote (since as $x \to -\infty$ $y \to 0$)
- For large positive values of x, y also takes large positive values (since as $x \to \infty$, $y \to \infty$)

> When x is negative,
> $$y = \frac{1}{e^{|x|}},$$
> so as x gets large and negative, y gets closer to zero.

Notice also if $f(x) = e^x$ then:

- The domain of f is the set of all real numbers, which can be written as $D(f) = (-\infty, \infty)$
- The range of f is the set of all positive real numbers, (i.e. $f(x) > 0$), which can be written as $R(f) = (0, \infty)$

Remember from your AS studies, that the statements $y = a^x$ and $x = \log_a y$ are equivalent power and logarithm versions of the same relationship.

Substituting e for a leads to the statements $y = e^x$ and $x = \log_e y$ being equivalent.

If $y = f(x) = e^x$, then to find the inverse we change the subject of the formula from y to x.

Remember, to find the inverse of a function, follow these steps:

Let the function equal y.

Rearrange the resulting equation to make x the subject.

Replace x with $f^{-1}(x)$ and replace y with x.

From the rules of logarithms we have $x = \log_e y$.

The logarithm to base e is also known as the natural logarithm and can be written as $\ln y$ instead of $\log_e y$.

So $x = \ln y$

Hence $f^{-1}(x) = \ln x$

Step by

The function f is defined by $f(x) = e^x$.

(a) Sketch the graph of $y = f(3x) - 4$, indicating the behaviour of your graph for large negative values of x.

(b) Write down the coordinates of the point of intersection of your graph with the y-axis.

(c) Find the x-coordinate of the point of intersection of the graph with the x-axis. Give your answer to three decimal places.

Steps to take

1   Do a sketch of the graph $y = e^x$, marking on the graph the coordinates of where the curve cuts the y-axis and also any asymptotes present.

2   Think about what effect each part of the new equation has on the curve $y = f(x)$.
$f(3x)$ is a stretch of scale factor $\left(\frac{1}{3}\right)$ parallel to the x-axis and the -4 represents a translation of $\begin{pmatrix} 0 \\ -4 \end{pmatrix}$.

3   Apply the two transformations to the first graph but draw it on a separate set of axes.

4   Mark on the coordinates of intersection with the y-axis.

5   The equation of the curve can be written as $y = e^{3x} - 4$. Along the x-axis, $y = 0$ so put $y = 0$ into the equation and solve the resulting equation for x.

· ·

Answer

(a)

When $x = 0$, $y = e^0 = 1$, so the curve cuts the x-axis at $(0, 1)$.

The x-axis is an asymptote for the curve.

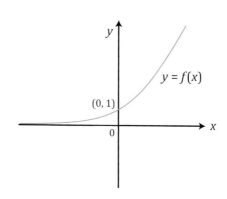

(b)

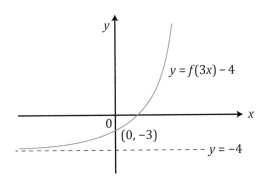

$y = f(3x) - 4$ cuts the y-axis at $(0, -3)$

> Note that the line $y = -4$ is an asymptote to the curve.

(c) At $y = 0$, $e^{3x} - 4 = 0$, giving $e^{3x} = 4$

Taking ln of both sides, we obtain $3x = \ln 4$

$$x = \tfrac{1}{3}\ln 4 = 0\ 462 \ (3 \text{ d.p.})$$

Example

1 The functions f and g have domains $[7, 60]$ and $[9, \infty)$ respectively and are defined by:

$$f(x) = 2\ln(4x + 5) + 3$$
$$g(x) = e^x.$$

(a) Find an expression for $f^{-1}(x)$.

(b) Write down the domain of f^{-1}, giving the end-points of your domain correct to the nearest integer.

(c) Write down an expression for $gf(x)$ and simplify your answer.

· ·

Answer

1 **(a)** Let $y = 2\ln(4x + 5) + 3$

$$y - 3 = 2\ln(4x + 5)$$
$$\frac{y - 3}{2} = \ln(4x + 5)$$

Taking exponentials of both sides, we obtain

$$e^{\frac{y-3}{2}} = 4x + 5$$
$$\frac{e^{\frac{y-3}{2}} - 5}{4} = x$$
$$f^{-1}(x) = \frac{e^{\frac{x-3}{2}} - 5}{4}$$

(b) When $x = 7$, $f(7) = 2\ln(4(7) + 5) + 3 = 9.99301... = 10$ (nearest integer)

When $x = 60$, $f(60) = 2\ln(4(60) + 5) + 3 = 14.00251...$
$$= 14 \text{ (nearest integer)}$$

$D(f^{-1}) = [10, 14]$

> Remember that the domain of f^{-1} equals the range of f, so to find the range of f the domain values are substituted for x in the function.
>
> Square brackets are used here as both of the end-points are allowable to the nearest integer.

(c) $gf(x) = e^{2\ln(4x+5)+3}$

$= e^{2\ln(4x+5)} \times e^3$

$= e^{\ln(4x+5)^2} \times e^3$

$= (4x+5)^2 \times e^3$

$= e^3(4x+5)^2$

The function ln x and its graph

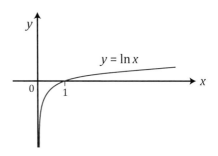

Notice that:

- The graph of $y = \ln x$ cuts the x-axis at $x = 1$. This is because when $y = 0$, $\ln x = 0$ and $\ln 1 = 0$.
- For large positive values of x, the y-values become large and positive (i.e. as $x \to \infty$, $y = \ln x \to \infty$)
- For small values of x, the y-values become large and negative (i.e. as $x \to \infty$, $y = \ln x \to -\infty$), so the y-axis is an asymptote.

Notice also if $f(x) = \ln x$:

- The domain of f is the set of all positive real numbers, i.e. $x > 0$, which can be written as $D(f) = (0, \infty)$
- The range of f is the set of all real numbers, i.e. $-\infty < x < \infty$, which can be written as $R(f) = (-\infty, \infty)$.

ln x as the inverse function of e^x

The inverse of a function will produce the input value from the output value. The inverse of $y = e^x$ is $y = \ln x$.

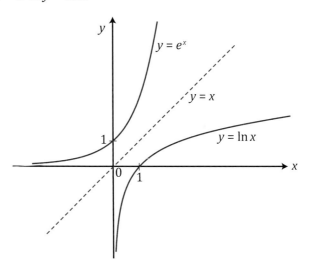

The graph of $y = \ln x$ can be obtained by reflecting the graph of $y = e^x$ in the line $y = x$.

Notice also that:

- The x-axis is an asymptote to the curve $y = e^x$.

- The y-axis is an asymptote to the curve $y = \ln x$.

- Since the action of an inverse function reverses the effect of the function, then two useful results follow: $e^{\ln x} = x$ and $\ln e^x = x$.

> When reflecting in the line $y = x$, you need to ensure that the scales on both axes are the same.

Examples

1 The function f has domain $(-\infty, \infty)$ and is defined by

$$f(x) = 3e^{2x}$$

The function g has domain $(0, \infty)$ and is defined by

$$g(x) = \ln 4x$$

(a) Write down the domain and range of fg.

(b) Solve the equation $fg(x) = 12$

. . . .

Answer

1 (a) Domain of fg is the domain of g, i.e. $(0, \infty)$

$$fg(x) = 3e^{2g(x)} = 3e^{2\ln 4x}$$

$$fg(x) = 3e^{\ln (4x)^2}$$

$$fg(x) = 3e^{\ln 16x^2}$$

$$fg(x) = 3(16x^2)$$

$$fg(x) = 48x^2$$

Now $D(fg) = (0, \infty)$

$fg(0) = 48(0)^2 = 0$ and as $x \to \infty, fg(x) \to \infty$

Hence $R(fg) = (0, \infty)$

> Note that $e^{\ln g(x)} = g(x)$

> The domain of a composite function fg is the domain of g.

> To find the range of fg, find the composite function and then, by substituting values from the domain, find its maximum and minimum values.

(b) $fg(x) = 48x^2$

Now $\qquad fg(x) = 12$

So $\qquad 48x^2 = 12$

$$x^2 = \frac{1}{4}$$

$$x = \pm\frac{1}{2}$$

However as the domain is $(0, \infty)$, $x \ne -\frac{1}{2}$

Hence solution is $x = \frac{1}{2}$

> It is important to check these values against the domain of the composite function to see if both values are allowable.

2   Given that $f(x) = \ln x$, sketch the graphs of $y = f(x)$ and $y = -f(x + 1)$ on the same diagram. Label the coordinates of the points of intersection with the x-axis and indicate the behaviour of the graphs for large positive and negative values of y.

Answer

2

You need to be able to reproduce the graph of $y = \ln x$, showing the point of intersection with the x-axis ($\ln 1 = 0$) and the y-axis as an asymptote.

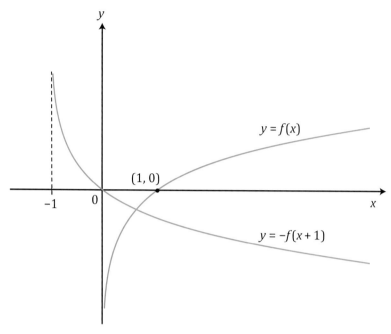

(Note the above graph would be hand-drawn in the exam.)

For $y = f(x) = \ln x$

The curve passes through the point $(1, 0)$, since $\ln 1 = 0$.

For large positive values of x, there are large positive values of y.

For negative values of x, the function $f(x) = \ln x$ does not exist.

For values of x between 0 and 1, $f(x) = \ln x$ is negative.

As x approaches 0 from above, the negative values of y become larger, (i.e. as $x \to 0$, $\ln x \to -\infty$), so the y-axis is an asymptote.

$y = -f(x + 1)$ represents a translation of $y = f(x)$ by one unit to the left followed by reflection in the x-axis.

For $y = -f(x + 1)$

The curve passes through the point $(0, 0)$.

The line $x = -1$ is an asymptote.

3   Given that $f(x) = e^x$, sketch on the same diagram the graphs of $y = f(x)$ and $y = -f(x) + 1$. Label the coordinates of the points of intersection of each of the graphs with the axes. Indicate the behaviour of each of the graphs for large positive and negative values of x.

Answer

3   (Note that in the exam this graph would be hand-drawn.)

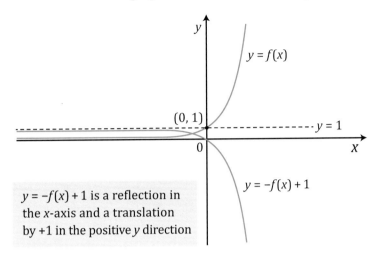

$y = -f(x) + 1$ is a reflection in the x-axis and a translation by +1 in the positive y direction

$y = f(x) = e^x$ cuts the y-axis at $(0, 1)$, since $e^0 = 1$.

The curve has an asymptote of $y = 0$ (or the x-axis), since as $x \rightarrow -\infty$,

$e^x \rightarrow 0$.

When $x \rightarrow +\infty$, the y-values also become large and positive (i.e. $y \rightarrow +\infty$).

$y = -f(x) + 1$ cuts the y-axis at $(0, 0)$ and has an asymptote of $y = 1$ (i.e. as $x \rightarrow -\infty$, $y \rightarrow 1$).

When $x \rightarrow +\infty$, the y-values become large and negative (i.e. $y \rightarrow -\infty$).

BOOST

Grade ⇧⇧⇧⇧

Make sure you have labelled the curve, any points of intersection asked for in the question, any asymptotes and the axes.

4   The function f has domain $(-\infty, \infty)$ and is defined by

$f(x) = 3e^{2x}$

The function g has domain $[1, \infty)$ and is defined by

$g(x) = 2 \ln x$

(a) Explain why $gf(-1)$ does not exist.

(b) Find in its simplest form an expression for $fg(x)$.

Answer

4   (a) $f(-1) = 3e^{-2} < 1$ so that $f(-1)$ is not in the domain of g, i.e. not in $[1, \infty)$.
        Hence $gf(-1)$ does not exist.

(b) $fg(x) = 3e^{2g(x)}$

        $= 3e^{2(2 \ln x)}$

        $= 3e^{4 \ln x}$

        $= 3e^{\ln x^4}$

        $= 3x^4$

Use this law of logarithms:
$k \log_a x = \log_a x^k$

Use $e^{\ln a} = a$

In order to be able to work out ranges when a function is given you must be familiar with the shapes of the graphs of the functions for given domains. You also need to know the shapes of curves to produce transformations on them.

Produce a crib sheet showing the shapes of the following curves:

$$f(x) = 2x - 3$$
$$f(x) = (x - 3)(x + 1)$$
$$f(x) = e^x$$
$$f(x) = e^x - 2$$
$$f(x) = \ln(x)$$

Feel free to add functions of your own and remember to check them using Google.

Check the shapes of the curve by typing in the equation of the curve into Google.

Use of functions in modelling

Functions can be used to model situations in the real world. For example, they can be used to model exponential decay (e.g. radioactive decay), exponential growth (e.g. the number of algae in a pond during hot weather) or the behaviour of tides.

Example

1   The rise and fall of the tide in a harbour with time can be modelled using the function

$$f(t) = 3.2 \sin(2.7t + 8.5) \quad \text{where the angle is measured in radians,}$$

where $f(t)$ is the vertical height of the water above or below the mean sea level in metres and t is the time in hours.

(a)   Find the height of the tide above mean sea level when $t = 0$. Give your answer to 2 significant figures.

(b)   State with a reason what the maximum height of the tide above sea level is.

Answer

1   (a)   When $t = 0$, $f(0) = 3.2 \sin(2.7 \times 0 + 8.5)$

$$= 3.2 \sin 8.5$$

$$= 2.6 \text{ m}$$

(b)   The maximum value of $\sin(2.7t + 8.5)$ is 1. Hence the maximum height above sea level $= 3.2 \times 1 = 3.2$ m

Test yourself

1 The function f has domain $x \le -2$ and is defined by $f(x) = (x + 2)^2 - 1$.
   (a) Find the range of f.   [2]
   (b) Find an expression for $f^{-1}(x)$. State the domain and range of f^{-1}.   [3]

2 The diagram shows a sketch of the graph of $y = f(x)$. The graph has its highest point at $(3, 4)$ and intersects the x-axis at the points $(1, 0)$ and $(5, 0)$.

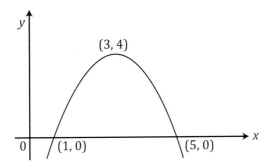

Sketch the graph of $y = 2f(x - 1)$, indicating the coordinates of three points on the graph.   [4]

3 Solve the following
   (a) $3|x - 1| + 7 = 19$ [2]
   (b) $6|x| - 3 = 2|x| + 5$ [2]

4 (a) Sketch the graph of $x^2 + 6x + 13$, identifying the stationary point.   [2]
   (b) The function f is defined by $f(x) = x^2 + 6x + 13$ with domain (a, b).
      (i) Explain why f^{-1} does not exist when $a = -10$ and $b = 10$.   [1]
      (ii) Write down a value of a and a value of b for which the inverse of f does exist and derive an expression for $f^{-1}(x)$.   [5]

5 The function f has domain $(-\infty, 10)$ and is defined by:

$$f(x) = e^{5 - \frac{x}{2}} + 6$$

   (a) Find an expression for $f^{-1}(x)$.   [4]
   (b) Write down the domain of f^{-1}.   [2]

6 Given that $f(x) = e^x$, sketch on the same diagram, the graphs of $y = f(x)$ and $y = 2f(x) - 2$. Label the coordinates of the points of intersection of each of the graphs with the x-axis. Indicate the behaviour of each of the graphs for large positive and negative values of x.   [6]

7 The function f has domain $(-\infty, \infty)$ and is defined by
       $f(x) = 2e^{3x}$
   The function g has domain $(0, \infty)$ and is defined by
       $g(x) = \ln 2x$

   (a) Write down the domain and range of fg.   [2]
   (b) Solve the equation $fg(x) = 128$.   [3]

Summary

Check you know the following facts:

Partial fractions

$$\frac{\alpha x + \beta}{(cx + d)(ex + f)} \equiv \frac{A}{cx + d} + \frac{B}{ex + f} \qquad\qquad \text{(i)}$$

$$\frac{\alpha x^2 + \beta x + \gamma}{(cx + d)(ex + f)^2} \equiv \frac{A}{cx + d} + \frac{B}{ex + f} + \frac{C}{(ex + f)^2} \qquad \text{(ii)}$$

In both cases, clear the fractions and choose appropriate values of x.
In (ii), an equation involving coefficients of x^2 may be used.

Functions

A function is a relation between a set of inputs and a set of outputs such that each input is related to exactly one output.

The domain and range of a function

The domain is the set of input values that can be entered into a function.
The range is the set of output values from a function.

Composition of functions

Composition of functions involves applying two or more functions in succession. The composite function $fg(x)$ means $f(g(x))$ and is the result of performing the function g first and then f.

One-to-one functions

A function where one output value would correspond to only one possible input value.

To find $f^{-1}(x)$ given $f(x)$

First check that $f(x)$ is a one-to-one function. Let y equal the function and rearrange so that x is the subject of the equation. Replace x on the left with $f^{-1}(x)$ and on the right replace all occurrences of y with x.

Domain and range of inverse functions

The range of $f^{-1}(x)$ is the same as the domain of $f(x)$.

The domain of $f^{-1}(x)$ is the same as the range of $f(x)$.

Graphs of inverse functions

The graph of $y = f^{-1}(x)$ is obtained by reflecting the graph of $y = f(x)$ in the line $y = x$.

The modulus function

The modulus of x is written $|x|$ and means the numerical value of x (ignoring the sign).

So whether x is positive or negative, $|x|$ is always positive (or zero).

Graphs of modulus functions

First plot the graph of $y = f(x)$ and reflect any part of the graph below the x-axis in the x-axis. The resulting graph will be $y = |f(x)|$.

Combinations of transformations

If a graph of $y = f(x)$ is drawn, then the graph of $y = f(x - a) + b$ can be obtained by applying the translation $\binom{a}{b}$ to the original graph.

If a graph of $y = f(x)$ is drawn, then the graph of $y = af(x - b)$ can be obtained by applying the following two transformations in either order: a stretch parallel to the y-axis with scale factor a and a translation of $\binom{b}{0}$.

If a graph of $y = f(x)$ is drawn, then the graph of $y = f(ax)$ can be obtained by scaling the x values by $\frac{1}{a}$.

The function e^x and its graph

$y = e^x$

> e^x is a one-to-one function and has an inverse $\ln x$.
> The graph of $y = e^x$ cuts the y-axis at $y = 1$ and has the x-axis as an asymptote.

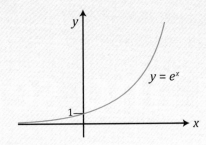

$D(f) = (-\infty, \infty)$
$R(f) = (0, \infty)$

The function $\ln x$ and its graph

$y = \ln x$

> $\ln x$ is a one-to-one function and has an inverse e^x.
> The graph of $y = \ln x$ cuts the x-axis at $x = 1$ and has the y-axis as an asymptote.

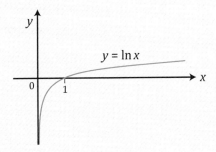

$D(f) = (0, \infty)$
$R(f) = (-\infty, \infty)$

The functions $y = \ln x$ and $y = e^x$ are inverse functions.

So, the graphs of $y = \ln x$ and $y = e^x$ are reflections of each other in the line $y = x$.

Also, $e^{\ln x} = x$ and $\ln e^x = x$.

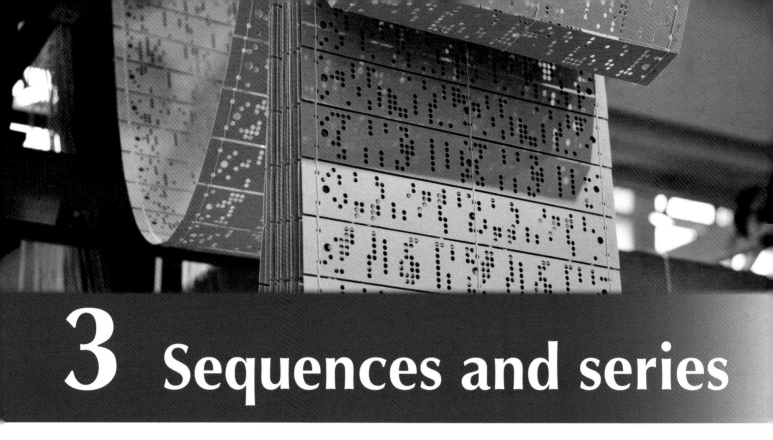

3 Sequences and series

Introduction

Some of this topic was covered in Topic 4 of the AS course so it would be best to have a quick look over this topic before starting this topic at A2 level.

In this topic you will be looking at sequences and series. Here you will learn how to expand linear expressions raised to different integer and fractional powers and how these expansions can be used to make simplifications. You will also learn about how to work out terms in series and how to work out the sum of the terms in a series.

This topic covers the following:

3.1 Binomial expansion for positive integral indices

3.2 The binomial expansion of $(a + bx)^n$ for any rational value of n

3.3 Using Pascal's triangle to work out the coefficients of the terms in the binomial expansion

3.4 The binomial expansion where $a = 1$

3.5 The binomial theorem for other values of n

3.6 The difference between a series and a sequence

3.7 Arithmetic sequences and series

3.8 Proof of the formula for the sum of an arithmetic series

3.9 The summation sign and its use

3.10 Geometric sequences and series

3.11 The difference between a convergent and a divergent sequence and series

3.12 Proof of the formula for the sum of a geometric series

3.13 The sum to infinity of a convergent geometric series

3.14 Sequences generated by a simple recurrence relation of the form $x_{n+1} = f(x_n)$

3.15 Increasing sequences

3.16 Decreasing sequences

3.17 Periodic sequences

3.18 Using sequences and series in modelling

3.1 Binomial expansion for positive integral indices

The binomial expansion was covered in AS Unit 1, so if you are unsure on the basics you should look back at Topic 4 of the AS book.

There are several formulae you need to use. All these formulae are included in the formula booklet.

Binomial expansion is the expansion of the expression $(a + b)^n$ where n is a positive integer.

The formula for the expansion will be given in the formula booklet and is shown here:

$$(a + b)^n = a^n + \binom{n}{1}a^{n-1}b + \binom{n}{2}a^{n-2}b^2 + \ldots + \binom{n}{r}a^{n-r}b^r + \ldots + b^n$$

> You do not need to memorise these formulae as they are given in the formula booklet.

where,

$$\binom{n}{r} = {}^nC_r = \frac{n!}{r!(n-r)!}$$

$n!$ means n factorial. If $n = 5$, then $5! = 1 \times 2 \times 3 \times 4 \times 5$ which you may also see written as $1.2.3.4.5$

Note that $0! = 1$.

3.2 The binomial expansion of $(a + bx)^n$ for any rational value of n

You may be asked to find the expansion of an expression such as $(4 + 2x)^3$.

Here you would substitute $a = 4$, $b = 2x$ and $n = 3$ into the formula for the expansion of $(a + b)^n$.

It should be noted that the binomial expansion of an expression having the form:

$(a + bx)^n$, is valid for $\left|\frac{bx}{a}\right| < 1$.

This means the expansion of $(3 + 2x)^6$ would be valid for $\left|\frac{2x}{3}\right| < 1$, so $|x| < \frac{3}{2}$.

Example

1 Expand $(3 + 2x)^6$ in ascending powers of x up to and including the term in x^3.

Answer

1    $(a + b)^n = a^n + \binom{n}{1}a^{n-1}b + \binom{n}{2}a^{n-2}b^2 + \ldots + \binom{n}{r}a^{n-r}b^r + \ldots + b^n$

     $a = 3$, $b = 2x$ and $n = 6$

     $(3 + 2x)^6 = 3^6 + \binom{6}{1}(3^5)(2x) + \binom{6}{2}(3^4)(2x)^2 + \binom{6}{3}(3^3)(2x)^3 + \ldots$

           $= 729 + 2916x + 4860x^2 + 4320x^3$

BOOST

Grade ⇧⇧⇧⇧

You may think that some of the numbers are getting too large and you have done something wrong. The numbers in questions like this are often large or small so keep going.

3.3 Using Pascal's triangle to work out the coefficients of the terms in the binomial expansion

You can also find the coefficients in the expansion of $(a + b)^n$ by using Pascal's triangle.

Suppose you want to expand the expression from the previous example, $(a + b)^4$ using Pascal's triangle.

You would write down Pascal's triangle and look for the line starting 1 and then 4 (because n is 4 here). The line 1, 4, 6, 4, 1 gives the coefficients. This avoids the calculation involving the factorials for each coefficient but you will have to remember how to construct Pascal's triangle.

$$1$$
$$1 \quad 1$$
$$1 \quad 2 \quad 1$$
$$1 \quad 3 \quad 3 \quad 1$$
$$1 \quad 4 \quad 6 \quad 4 \quad 1$$
$$1 \quad 5 \quad 10 \quad 10 \quad 5 \quad 1$$

Hence inserting the coefficients 1, 4, 6, 4, 1 we have the following expansion:

$$(a + b)^4 = a^4 + 4a^3b + 6a^2b^2 + 4ab^3 + b^4$$

3.4 The binomial expansion where $a = 1$

When the first term in the bracket (i.e. a) is 1 and b is replaced by x, the binomial expansion when n is a positive integer, becomes:

$$(1 + x)^n = 1 + nx + \frac{n(n - 1)x^2}{2!} + \frac{n(n - 1)(n - 2)x^3}{3!} + \dots$$

Again this formula is given in the formula booklet so you don't need to memorise it.

3.5 The binomial theorem for other values of n

In the previous section, which referred to the binomial expansion covered in AS Unit 1, it should be noted that the index n was a positive integer.

However, the expansion

$$(1 + x)^n = 1 + nx + \frac{n(n - 1)x^2}{2!} + \frac{n(n - 1)(n - 2)x^3}{3!} + \dots$$

holds for negative or fractional values of n, **provided that x lies between ±1, i.e. $|x| < 1$.**

For this range of values of x the series is convergent; because as x is raised to increasing powers, the values of successive terms decrease rapidly. As a result, once the first few terms have been added, subsequent terms become small and insignificant and the sum of the series approaches a final steady value. If $|x| < 1$ did not apply, the subsequent terms would get larger and larger and it would be impossible to determine an approximation to the series.

Examples

Here are the values of x for which the expansions of each of the following are convergent.

1 $(1 + x)^{-1}$ expansion is convergent for $|x| < 1$.

2 $\left(1 + \dfrac{x}{3}\right)^{-1}$ $\left|\dfrac{x}{3}\right| < 1$, so expansion is convergent for $|x| < 3$.

3 $(1 - x)^{-1}$ expansion is convergent for $|x| < 1$.

The condition for convergence when two binomial expansions are combined

Suppose the following two binomial expansions are added together:

$$(1 + 2x)^5 + \left(1 - \frac{x}{2}\right)^4$$

$(1 + 2x)^5$ We require $|2x| < 1$, so this expansion is convergent for $|x| < \dfrac{1}{2}$.

$\left(1 - \dfrac{x}{2}\right)^4$ We require $\left|\dfrac{x}{2}\right| < 1$, so this expansion is convergent for $|x| < 2$.

Now the value of x has to satisfy both conditions and as $|x| < \dfrac{1}{2}$ lies inside $|x| < 2$, the condition for the combined convergence is $|x| < \dfrac{1}{2}$ or $-\dfrac{1}{2} < x < \dfrac{1}{2}$.

Examples

1   Expand $(2 - x)^{\frac{1}{2}}$ in ascending powers of x up to and including the term in x^2.

    State the range of values of x for which the expansion is valid.

. . . .

Answer

1 $(2 - x)^{\frac{1}{2}} = \left[2\left(1 - \dfrac{x}{2}\right)\right]^{\frac{1}{2}}$

> Notice how the 2 is removed out of the bracket. It is important to note that this 2 is raised to the power $\frac{1}{2}$.

    $= 2^{\frac{1}{2}}\left(1 - \dfrac{x}{2}\right)^{\frac{1}{2}}$

    $= \sqrt{2}\left[1 + \left(\dfrac{1}{2}\right)\left(\dfrac{-x}{2}\right) + \dfrac{\left(\dfrac{1}{2}\right)\left(-\dfrac{1}{2}\right)\left(\dfrac{-x}{2}\right)^2}{2!} + \ldots\right]$

> Use the formula for the binomial expansion:
> $$(1 + x)^n = 1 + nx + \frac{n(n - 1)x^2}{2!} + \ldots$$
> obtained from the formula booklet with $n = \frac{1}{2}$ and x replaced by $-\frac{x}{2}$.

    $= \sqrt{2}\left(1 - \dfrac{x}{4} - \dfrac{x^2}{32} + \ldots\right)$

For the expansion of $\left(1 - \dfrac{x}{2}\right)^{\frac{1}{2}}$ to converge, we require $\left|\dfrac{x}{2}\right| < 1$,

i.e. $|x| < 2$ or $-2 < x < 2$.

2   Expand $\left(1 - \dfrac{x}{4}\right)^{\frac{1}{2}}$ in ascending powers of x up to and including the term in x^2.

State the range of values of x for which your expansion is valid.

Hence, by writing $x = 1$ in your expansion, show that
$$\sqrt{3} = \frac{111}{64}$$

Answer

> *This formula for the binomial expansion is given in the formula booklet so you don't need to memorise it.*

2   Now $(1 + x)^n = 1 + nx + \dfrac{n(n-1)x^2}{2!} + \dfrac{n(n-1)(n-2)x^3}{3!} + \dots$

Here $n = \dfrac{1}{2}$ and x is replaced with $-\dfrac{x}{4}$, so

$$\left(1 - \frac{x}{4}\right)^{\frac{1}{2}} = 1 + \left(\frac{1}{2}\right)\left(-\frac{x}{4}\right) + \frac{\left(\frac{1}{2}\right)\left(-\frac{1}{2}\right)}{1 \times 2}\left(\frac{x^2}{16}\right) + \dots$$

> *Notice the +... here. It means that the series continues.*

$$= 1 - \frac{x}{8} - \frac{x^2}{128} + \dots$$

> *This is shown in the formula booklet.*

When n is negative or fractional, the $(1 + x)^n$ expansion is valid for $|x| < 1$.

Then for $\left(1 - \dfrac{x}{4}\right)^{\frac{1}{2}}$, the expansion is valid for $\left|\dfrac{x}{4}\right| < 1$, or $|x| < 4$ i.e. $-4 < x < 4$.

When $x = 1$, $\left(1 - \dfrac{x}{4}\right)^{\frac{1}{2}} = \left(1 - \dfrac{1}{4}\right)^{\frac{1}{2}} = \left(\dfrac{3}{4}\right)^{\frac{1}{2}} = \dfrac{\sqrt{3}}{2}$

> *The approximately equals sign is used now because the series is only approximate to the expansion as only the first three terms are used.*

Hence

$$\frac{\sqrt{3}}{2} \approx 1 - \frac{1}{8} - \frac{1}{128} \approx \frac{111}{128}$$

$$\frac{\sqrt{3}}{2} \approx \frac{111}{128}$$

$$\sqrt{3} \approx \frac{111}{64}$$

---

3   **(a)**   Expand $\dfrac{(1 + 3x)^{\frac{1}{3}}}{(1 + 2x)^2}$ in ascending powers of x up to and including the term in x^2.

State the range of values of x for which your expansion is valid.

(b)   Use your expansion to find an approximate non-zero value of x satisfying the equation
$$\frac{(1 + 3x)^{\frac{1}{3}}}{(1 + 2x)^2} = 1 - 4x - 2x^2.$$

Answer

3   **(a)** $\dfrac{(1 + 3x)^{\frac{1}{3}}}{(1 + 2x)^2} = (1 + 3x)^{\frac{1}{3}}(1 + 2x)^{-2}$

$$= \left[1 + \left(\frac{1}{3}\right)(3x) + \frac{\left(\frac{1}{3}\right)\left(-\frac{2}{3}\right)(3x)^2}{1.2} + \dots\right] \times \left[1 + (-2)(2x) + \frac{(-2)(-3)(2x)^2}{1.2} + \dots\right]$$

> *Note when multiplying out the bracket only terms up to and including x^2 are included.*

$$= [1 + x - x^2 + \dots][1 - 4x + 12x^2 + \dots]$$

$$= 1 - 4x + 12x^2 + x - 4x^2 - x^2 + \dots$$

$$= 1 - 3x + 7x^2 + ...$$

The expansion is valid for $|3x| < 1$ and $|2x| < 1$, i.e. valid for $|x| < \dfrac{1}{3}$

(b) Then replacing $\dfrac{(1 + 3x)^{\frac{1}{3}}}{(1 + 2x)^2}$ by $1 - 3x + 7x^2$

we have $1 - 3x + 7x^2 \approx 1 - 4x - 2x^2$

so that $\qquad 9x^2 = -x$

$$x(9x + 1) = 0$$

Hence $x = 0$ or $x = -\dfrac{1}{9}$

The required approximate non-zero value is $x = -\dfrac{1}{9}$ or -0.111 correct to 3 decimal places

---

4 (a) Expand $(4 - x)^{\frac{3}{2}}$ as far as the term in x^2.

(b) Use your result in part (a) to expand $\dfrac{(4 - x)^{\frac{3}{2}}}{(1 + 2x)}$ as far as the term in x^2.

State the range of values of x for which the expansion is valid.

· ·

Answer

4 (a) $(4 - x)^{\frac{3}{2}} = \left[4 \left(1 - \dfrac{x}{4} \right) \right]^{\frac{3}{2}} = 4^{\frac{3}{2}} \times \left(1 - \dfrac{x}{4} \right)^{\frac{3}{2}} = 8 \left(1 - \dfrac{x}{4} \right)^{\frac{3}{2}}$

Note that $4^{\frac{3}{2}} = \sqrt{4^3} = 8$

Using the formula for the binomial expansion:

$$(1 + x)^n = 1 + nx + \frac{n(n - 1)x^2}{2!} + \frac{n(n - 1)(n - 2)x^3}{3!} + ...$$

This is obtained from the formula booklet.

$$8 \left(1 - \frac{x}{4} \right)^{\frac{3}{2}} = 8 \left[1 + \left(\frac{3}{2} \right) \left(-\frac{x}{4} \right) + \frac{\left(\frac{3}{2} \right) \left(\frac{1}{2} \right) \left(-\frac{x}{4} \right)^2}{2!} + ... \right]$$

$$= 8 \left[1 - \frac{3x}{8} - \frac{3x^2}{128} + ... \right]$$

$$= 8 - 3x + \frac{3x^2}{16} + ...$$

(b) $(1 + 2x)^{-1} = 1 + (-1)(2x) + \dfrac{(-1)(-2)(2x)^2}{2!} + ...$

$$= 1 - 2x + 4x^2 + ...$$

Note when multiplying out the bracket only terms up to and including x^2 are included.

Hence $\dfrac{(4 - x)^{\frac{3}{2}}}{(1 + 2x)} = \left(1 - 2x + 4x^2 + ... \right) \left(8 - 3x + \dfrac{3x^2}{16} + ... \right)$

$$= 8 - 3x + \frac{3x^2}{16} - 16x + 6x^2 + 32x^2 + ...$$

$$= 8 - 19x + \frac{611}{16} x^2 + ...$$

$\left(1 - \dfrac{x}{4} \right)^{\frac{3}{2}}$ is valid for $\left| \dfrac{x}{4} \right| < 1$, so $|x| < 4$ or $-4 < x < 4$

Notice that the second range is inside the first range and as x has to be valid for both the expansions, the expansions are valid for
$$|x| < \frac{1}{2}, \text{ or } -\frac{1}{2} < x < \frac{1}{2}$$

$(1 + 2x)^{-1}$ is valid for $|2x| < 1$, so $|x| < \frac{1}{2}$ or $-\frac{1}{2} < x < \frac{1}{2}$

Hence $\dfrac{(4 - x)^{\frac{3}{2}}}{(1 + 2x)}$ is valid for $|x| < \frac{1}{2}$, or $-\frac{1}{2} < x < \frac{1}{2}$

---

5 Expand $\left(1 + \dfrac{x}{8}\right)^{-\frac{1}{2}}$ in ascending powers of x up to and including the term in x^2.

State the range of values of x for which your expansion is valid.

Hence, by writing $x = 1$ in your expansion, find an approximate value for $\sqrt{2}$ in the form $\dfrac{a}{b}$, where a and b are integers whose values are to be found.

. . . .

Answer

5 $\left(1 + \dfrac{x}{8}\right)^{-\frac{1}{2}} = 1 + \left(-\dfrac{1}{2}\right)\dfrac{x}{8} + \dfrac{\left(-\frac{1}{2}\right)\left(-\frac{3}{2}\right)\left(\frac{x}{8}\right)^2}{2} + \ldots = 1 - \dfrac{x}{16} + \dfrac{3x^2}{512} + \ldots$

Note that instead of using the modulus sign, this can be written as $-8 < x < 8$

Expansion is valid for $\left|\dfrac{x}{8}\right| < 1$ so $|x| < 8$

When $x = 1$, $\left(\dfrac{9}{8}\right)^{-\frac{1}{2}} \approx 1 - \dfrac{x}{16} + \dfrac{3x^2}{512}$

Now $\left(\dfrac{9}{8}\right)^{-\frac{1}{2}} = \sqrt{\dfrac{8}{9}} = \dfrac{1}{3}\sqrt{8} = \dfrac{2\sqrt{2}}{3}$

Hence $\dfrac{2\sqrt{2}}{3} \approx 1 - \dfrac{1}{16} + \dfrac{3}{512}$

$\sqrt{2} \approx \dfrac{3}{2}\left(1 - \dfrac{1}{16} + \dfrac{3}{512}\right)$

$\approx \dfrac{1449}{1024}$

a is 1449 and *b* is 1024

Step by STEP

Find an approximate value for one root of the equation
$$2(1 + 6x)^{\frac{1}{3}} = 2x^2 - 15x$$

Steps to take

1 Expand $(1 + 6x)^{\frac{1}{3}}$ using binomial expansion as far as the term in x^2. We only need to go as far as x^2 as this is the highest power of x on the right-hand side of the equation.

2 Multiply the expansion by 2.

3 Group all the terms on one side of the equation and then solve the resulting equation.

4 Check that each value of x is allowable given the range of values for which the binomial expansion is valid.

Answer

$$(1 + x)^n = 1 + nx + \frac{n(n-1)x^2}{2!} + \frac{n(n-1)(n-2)x^3}{3!} + \ldots$$

$$(1 + 6x)^{\frac{1}{3}} = 1 + \left(\frac{1}{3}\right)(6x) + \frac{\left(\frac{1}{3}\right)\left(-\frac{2}{3}\right)(6x)^2}{2} + \ldots$$

$$= 1 + 2x - 4x^2 + \ldots$$

Hence

$$2(1 + 2x - 4x^2) = 2x^2 - 15x$$

$$2 + 4x - 8x^2 = 2x^2 - 15x$$

$$10x^2 - 19x - 2 = 0$$

$$(10x + 1)(x - 2) = 0$$

Hence $x = -\frac{1}{10}$ or 2.

Now the binomial expansion is valid for $|6x| < 1$ or $|x| < \frac{1}{6}$ which means $-\frac{1}{6} < x < \frac{1}{6}$.

Hence $x = 2$ is outside this range and is ignored.

So approximate root is $x = -\frac{1}{10}$

3.6 The difference between a series and a sequence

A sequence is simply a list of terms (e.g. 1, 3, 5, 7, ...). A series is the sum of a certain number of terms of a sequence (e.g. 1 + 3 +5 + 7 is a series formed by the first 4 terms of the sequence).

3.7 Arithmetic sequences and series

In an arithmetic sequence, successive terms have a common difference d between them.

Take the following sequence, for example: 2, 5, 8, 11, ...

The above sequence has a first term of 2 and a common difference of 3. This common difference can be found by taking any term from the second term onwards and then subtracting the preceding term.

If the sequence starts with the first term a, then it carries on like this:

	$a,$	$a + d,$	$a + 2d,$	$a + 3d,$	...
Term:	1st	2nd	3rd	4th	

The terms of an arithmetic sequence can be written in the following way:

$$t_1 = a, \quad t_2 = a + d, \quad t_3 = a + 2d, \quad \text{etc.}$$

From the pattern in the terms you can see that the nth term,

$$t_n = a + (n - 1)d$$

3.8 Proof of the formula for the sum of an arithmetic series

An arithmetic series is formed when the terms of an arithmetic sequence are added together.

The sum of n terms of an arithmetic series can be written as:

$$S_n = a + (a + d) + (a + 2d) + \dots + (a + (n - 1)d) \qquad (1)$$

The above sum starts from the first term and adds successive terms until the last term.

Now reversing the sum of the series starting from the last term, which we can call l, gives:

$$S_n = l + (l - d) + (l - 2d) + \dots + (l - (n - 1)d) \qquad (2)$$

Adding (1) and (2) together gives:

$$2S_n = (a + l) + (a + l) + (a + l) + \dots + (a + l)$$

Notice that in the above, the $(a + l)$ appears n times.

Hence, we can write:

$$2S_n = n(a + l)$$

$$S_n = \frac{n}{2}(a + l)$$

The last term l can be written as $l = a + (n - 1)d$

So

$$S_n = \frac{n}{2}(a + a + (n - 1)d)$$

$$\boxed{S_n = \frac{n}{2}[2a + (n - 1)d]}$$

The above formula appears in the formula booklet.

> You must remember this proof as in the examination you may be asked to prove the formula:
> $$S_n = \frac{n}{2}[2a + (n - 1)d]$$
> The formula for the sum of an arithmetic series,
> $$S_n = \frac{n}{2}(a + l)$$
> can be a useful version of this formula at times.

Examples

1 Find the sum of the first 20 terms of the arithmetic series which starts:

$$4 + 11 + 18 + 25 + \dots$$

. . . .

Answer

1   First term $a = 4$ and common difference $d = 11 - 4 = 7$

$$S_n = \frac{n}{2}[2a + (n - 1)d]$$

$$S_{20} = \frac{20}{2}[2 \times 4 + (20 - 1)7]$$

$$S_{20} = 1410$$

> This formula can be obtained from the formula booklet.

---

2 **(a)** An arithmetic series has first term a and common difference d. Prove that the sum of the first n terms of the series is given by

$$S_n = \frac{n}{2}[2a + (n - 1)d]$$

(b) The eighth term of an arithmetic series is 28. The sum of the first 20 terms of the series is 710. Find the first term and the common difference of the arithmetic series.

(c) The first term of another arithmetic series is –3 and fifteenth term is 67. Find the sum of the first fifteen terms of this arithmetic series.

Answer

2 (a) See Proof of the formula for the sum of an arithmetic series for this answer on page 59.

> Many of the exam questions in this topic require you to create equations using the information given in the question and then solve them simultaneously to find a and d.

(b)
$$t_n = a + (n-1)d$$
$$t_8 = a + 7d$$
$$28 = a + 7d \qquad (1)$$

$$S_n = \frac{n}{2}[2a + (n-1)d]$$
$$S_{20} = \frac{20}{2}[2a + 19d]$$
$$710 = 10(2a + 19d)$$
$$71 = 2a + 19d \qquad (2)$$

> Both sides are divided by 10.

Equations (1) and (2) are solved simultaneously.

$$71 = 2a + 19d$$
$$56 = 2a + 14d$$

Subtracting $\overline{\quad 15 = 5d \quad}$

> Equation (1) is multiplied by 2 before subtracting.

Giving $d = 3$

Substituting $d = 3$ into equation (1) gives $28 = a + 21$

Giving $a = 7$

Hence common difference = 3 and first term = 7

(c)
$$t_n = a + (n-1)d$$
15th term $\quad = -3 + 14d$
$$67 = -3 + 14d$$

Solving gives $d = 5$

$$S_n = \frac{n}{2}[2a + (n-1)d]$$

$$S_{15} = \frac{15}{2}[-6 + 14 \times 5] = 480$$

> By using the formula in the form
> $$S_n = \frac{n}{2}(a + l)$$
> the answer can be found more readily, i.e.
> $$S_{15} = \frac{15}{2}(-3 + 67) = 480$$

3.9 The summation sign and its use

If you wanted to add the first four terms of an arithmetic sequence to form an arithmetic series, you can write it like this: $t_1 + t_2 + t_3 + t_4$.

This can be written in the following way using a summation sign Σ.

$$\sum_{n=1}^{4} t_n$$

This means the sum of the terms from $n = 1$ to 4.

Take the following example. Here the terms are found by substituting $n = 1$, $n = 2$, $n = 3$, $n = 4$ and $n = 5$ into $(2n + 3)$. The terms are added together to form the series.

$$\sum_{n=1}^{5} (2n + 3) = 5 + 7 + 9 + 11 + 13 = 45$$

Examples

1   Evaluate $\displaystyle\sum_{n=1}^{3} n(n + 1)$.

Answer

1   $\displaystyle\sum_{n=1}^{3} n(n + 1) = 1 \times 2 + 2 \times 3 + 3 \times 4 = 2 + 6 + 12 = 20$

---

2   Evaluate $\displaystyle\sum_{n=1}^{100} (2n - 1)$.

Answer

2   Series is $1 + 3 + 5 + 7 + 9\ldots$

> Start off by writing the first few terms so that a and d can be found.

First term $a = 1$ and common difference $d = 2$.

> This formula is obtained from the formula booklet.

$$S_n = \frac{n}{2}\big[2a + (n - 1)d\big]$$

$$S_{100} = \frac{100}{2}\big[2 \times 1 + (100 - 1)2\big]$$

> Remember to do the multiplication before the addition in the square bracket.

$$S_{100} = 50\big[2 + 99 \times 2\big]$$

$$S_{100} = 10\,000$$

---

3   The nth term of a number sequence is denoted by t_n. The $(n + 1)$th term of the sequence satisfies $t_{n+1} = 2t_n - 3$ for all positive integers n, and $t_4 = 33$.

(a)   Evaluate t_1.

(b)   Explain why 40 098 cannot be one of the terms of this number sequence.

Answer

> Work backwards by substituting t_4 into the equation to find t_3. Repeat by substituting t_3 in to find t_2. Finally substitute t_2 in to find the answer t_1.

3   (a) $t_{n+1} = 2t_n - 3$

$t_4 = 2t_3 - 3$

$33 = 2t_3 - 3$

Solving gives $t_3 = 18$

$t_3 = 2t_2 - 3$

$18 = 2t_2 - 3$

Solving gives $t_2 = \dfrac{21}{2}$

$t_2 = 2t_1 - 3$

$\dfrac{21}{2} = 2t_1 - 3$

Solving gives $t_1 = \dfrac{27}{4}$

(b) $t_{n+1} = 2t_n - 3$

Doubling t_n will always result in an even number when t_n is a whole number (i.e. from t_3 onwards). Subtracting 3 from an even number always results in an odd number. The number 40 098 is even and therefore cannot be a term of the sequence.

> Look carefully at the equation to see what would happen if t_n was odd or even.

3.10 Geometric sequences and series

Here is an example of a geometric sequence: 1, 5, 25, 125, ...

From the second term onwards, if you divide one term by the term in front, you get the same number, which is called the common ratio.

In this series the common ratio is $\frac{25}{5} = 5$.

If the first term is a and the common ratio is r then a geometric sequence can be written as: $a, ar, ar^2, ar^3, ...$

Hence the first term $t_1 = a$, the second term $t_2 = ar$, the third term $t_3 = ar^2$, etc. Notice that the power of r is one less than the term number.

You can see that the | nth term, $t_n = ar^{n-1}$ |

The common ratio is found by dividing the second term onwards by its preceding term.

Hence, $\frac{t_2}{t_1} = \frac{ar}{a} = r, \quad \frac{t_3}{t_2} = \frac{ar^2}{ar} = r, \quad$ etc.

3.11 The difference between a convergent and a divergent sequence and series

A **convergent sequence** is a sequence with a limit that is a real number. In other words the sequence approaches a certain value.

The convergent sequence 2.1, 2.01, 2.001, 2.0001, ... approaches a value of 2.

A **divergent sequence** is a sequence that has a limit of infinity. The terms of the sequence keep getting larger and larger.

The divergent sequence 1, 2, 3, 4, ... will eventually reach infinity, ∞.

> Note that u_n means the nth term.

A **convergent series** is a series where the nth term gets closer to a certain number L as n approaches infinity. So, as $n \to \infty$, $u_n \to L$

This can also be written as $u_{n+1} = f(u_n)$ and $f(L)$.

A **divergent series** is a series where the nth term does not reach a steady value.

> This means that once you have reached a value L when this is put back into the function it will also give L.

Example

1   A sequence is generated using the following relation

$$a_n = \frac{n}{n+1}$$

(a)   Prove that the sequence generated is a convergent sequence.

(b)   The sequence eventually reaches a steady value. State this value.

Answer

1 (a) $n = 1$, $a_1 = \dfrac{1}{1+1} = \dfrac{1}{2}$

$n = 10$, $a_{10} = \dfrac{10}{10+1} = 0.9090 \ldots$

$n = 100$, $a_{100} = \dfrac{100}{100+1} = 0.9900 \ldots$

$n = 1000$, $a_{1000} = \dfrac{1000}{1000+1} = 0.9990 \ldots$

(b) As n approaches ∞, the fraction $\dfrac{n}{n+1}$ gets closer to 1

Steady value = 1

> Note that adding 1 to infinity is still infinity.

3.12 Proof of the formula for the sum of a geometric series

A geometric series is found by adding successive terms of a geometric sequence:

$$a + ar + ar^2 + ar^3 + \ldots + ar^{n-1}$$

The sum of n terms of a geometric series can be written as:

$$S_n = a + ar + ar^2 + ar^3 + \ldots + ar^{n-1} \qquad (1)$$

Multiplying S_n by r gives

$$rS_n = ar + ar^2 + ar^3 + ar^4 + \ldots + ar^n \qquad (2)$$

Subtracting equation (2) from equation (1) gives:

$$S_n - rS_n = a - ar^n$$

$$S_n(1 - r) = a(1 - r^n)$$

> This formula can be obtained from the formula booklet.

$$\boxed{S_n = \dfrac{a(1 - r^n)}{1 - r} \quad \text{provided that } r \neq 1}$$

Example

1 The fifth term of a geometric sequence is 96 and the eighth term is 768. Find the common ratio and the first term.

Answer

$$t_5 = ar^4 = 96$$

$$t_8 = ar^7 = 768$$

> Notice that dividing the terms, a cancels leaving an expression just in r.

Dividing these two terms $\dfrac{ar^7}{ar^4} = r^3 = \dfrac{768}{96} = 8$

Hence $r^3 = 8$ so common ratio $r = 2$

$$ar^4 = 96$$

So $a(2)^4 = 96$

Giving first term $a = 6$

3.13 The sum to infinity of a convergent geometric series

The following series is convergent: $1 + \frac{1}{2} + \frac{1}{4} + \frac{1}{8} + \dots$

This means that successive terms are getting smaller in magnitude and that S_n approaches a certain limiting value. As $n \to \infty$, the sum of all the terms is called the sum to infinity, symbol S_∞. The sum to infinity of a geometric series is given by:

$$S_\infty = \frac{a}{1-r} \quad \text{provided that } |r| < 1$$

For a geometric series, $S_n = \frac{a(1-r^n)}{1-r} = \frac{a}{1-r} - \frac{ar^n}{1-r}$

If $|r| < 1$, r^n becomes very small as $n \to \infty$.

This means $\frac{ar^n}{1-r} \to 0$ as $n \to \infty$.

Hence $\qquad S_n \to \frac{a}{1-r}$ as $n \to \infty$.

If r takes a value not in the range $|r| < 1$, successive terms in the series become larger, so the series is divergent and a final limiting value is not reached. In this case S_∞ would not exist.

Examples

1 (a) Find the sum to infinity of the geometric series:

$$40 - 24 + 14.4 - \dots$$

   (b) Another geometric series has first term a and common ratio r. The fourth term of this geometric series is 8. The sum of the third, fourth and fifth terms of the series is 28.

   (i) Show that r satisfies the equation: $2r^2 - 5r + 2 = 0$

   (ii) Given that $|r| < 1$, find the value of r and the corresponding value of a.

Answer

1 (a) The terms of a geometric series are as follows:

$$a + ar + ar^2 + ar^3 + \dots + ar^{n-1}$$

Common ratio $r = \dfrac{\text{2nd term}}{\text{1st term}}$

$$r = \frac{ar}{a} = -\frac{24}{40} = -\frac{3}{5}$$

$$S_\infty = \frac{a}{1-r}$$

$$= \frac{40}{1 - \left(-\frac{3}{5}\right)}$$

$$= 25$$

The nth term of a geometric series = ar^{n-1}.

(b) (i)   4th term = ar^3

Hence $8 = ar^3$ so $a = \dfrac{8}{r^3}$

3rd term = ar^2

5th term = ar^4

Now $ar^2 + ar^3 + ar^4 = 28$

Substituting $a = \dfrac{8}{r^3}$ into this equation gives the following:

$$\dfrac{8}{r^3}r^2 + \dfrac{8}{r^3}r^3 + \dfrac{8}{r^3}r^4 = 28$$

$$\dfrac{8}{r} + 8 + 8r = 28$$

Multiplying both sides by r gives:

$$8 + 8r + 8r^2 = 28r$$

$$8r^2 - 20r + 8 = 0$$

Dividing through by 4 gives:

$$2r^2 - 5r + 2 = 0$$

(ii)   Factorising $2r^2 - 5r + 2 = 0$

$$(2r - 1)(r - 2) = 0$$

Solving gives $r = \dfrac{1}{2}$ or $r = 2$

It is given that $|r| < 1$

So $r = \dfrac{1}{2}$

$$a = \dfrac{8}{r^3} = \dfrac{8}{\left(\frac{1}{2}\right)^3} = \dfrac{8}{\left(\frac{1}{8}\right)} = 64$$

---

2   A geometric series has first term a and common ratio r. The sum of the first two terms of the geometric series is 7.2. The sum to infinity of the series is 20. Given that r is positive, find the values of r and a.

· ·

Answer

2   1st term = a, 2nd term = ar,

So $a + ar = 7.2$

$a(1 + r) = 7.2$

This formula for the sum to infinity can be obtained from the formula booklet.

$$S_\infty = \dfrac{a}{1 - r}$$

$$20 = \dfrac{a}{1 - r}$$

$$a = 20(1 - r)$$

Substituting $a = 20(1 - r)$ into $a(1 + r) = 7.2$

This gives $20(1 - r)(1 + r) = 7.2$

$$20(1 - r^2) = 7.2$$

$$r^2 = 0.64$$

$$r = \pm 0.8 \text{ but } r \text{ is positive, so } r = 0.8$$

Now

$$a(1 + r) = 7.2$$

$$a(1 + 0.8) = 7.2$$

$$a = 4$$

---

3 **(a)** The second term of a geometric series is 6 and the fifth term is 384.

    (i) Find the common ratio of the series.

    (ii) Find the sum of the first eight terms of the geometric series.

(b) The first term of another geometric series is 5 and the common ratio is 1.1.

    (i) The nth term of this series is 170, correct to the nearest integer. Find the value of n.

    (ii) Dafydd, who has been using his calculator to investigate various properties of this geometric series, claims that the sum to infinity of the series is 940. Explain why this result cannot possibly be correct.

• •

Answer

3 **(a)** **(i)**

$$t_2 = ar$$

So

$$6 = ar \tag{1}$$

$$t_5 = ar^4$$

So,

$$384 = ar^4 \tag{2}$$

Dividing equation (2) by equation (1) gives

$$\frac{384}{6} = \frac{ar^4}{ar}$$

$$64 = r^3$$

Giving $r = 4$

(ii) Substituting $r = 4$ into equation (1)

$$6 = 4a$$

So

$$a = \frac{3}{2}$$

> It is often easier to write the formula in the form
> $$S_n = \frac{a(r^n - 1)}{r - 1} \text{ when } r > 1,$$
> thus avoiding a negative numerator and denominator.

$$S_n = \frac{a(1 - r^n)}{1 - r}$$

$$S_8 = \frac{\frac{3}{2}\left(1 - 4^8\right)}{1 - 4}$$

$$= 32\,767.5$$

> This formula can be obtained from the formula booklet.

(b) **(i)** $t_n = ar^{n-1}$

Hence

$$170 = 5(1.1)^{n-1}$$

$$34 = (1.1)^{n-1}$$

> Equations involving powers like this are solved by taking logs of both sides.

See the previous topic if you are unsure about taking logs of both sides.

BOOST
Grade ⬆⬆⬆⬆

Always look back at the question to see if there are any conditions placed on the value you have found. Here it has to be an integer.

Taking $\log_{10}$ of both sides:

$$\log_{10} 34 = \log_{10} (1.1)^{n-1}$$

$$\log_{10} 34 = (n-1)\log_{10} 1.1$$

$$\frac{\log_{10} 34}{\log_{10} 1.1} = n - 1$$

$$36.9988 = n - 1$$

$$n = 37.9988$$

n has to be an integer, so $n = 38$

(ii) The common ratio is 1.1 . For a sum to infinity to exist $|r| < 1$ so in this case the sum to infinity does not exist.

3.14 Sequences generated by a simple recurrence relation of the form $x_{n+1} = f(x_n)$

The following relation is called a recurrence relation:

$$x_{n+1} = x_n^3 + \frac{1}{9}$$

This recurrence relation can be used to generate a sequence by substituting a starting value called x_0 into the relation to calculate the next term in the sequence, called x_1. The value of x_1 is then substituted for x_n into the relation to calculate x_2. The process is repeated until the desired number of terms of the sequence have been found.

Example

1 A sequence is generated using the recurrence relation

$$x_{n+1} = x_n^3 + \frac{1}{9}$$

Starting with $x_0 = 0.1$, find and record x_1, x_2, x_3 .

· ·

Answer

Do not round off any of your values. Write down the full calculator display.

1   $x_0 = 0.1$

$x_1 = x_0^3 + \dfrac{1}{9} = (0.1)^3 + \dfrac{1}{9} = 0.1121111111$

$x_2 = x_1^3 + \dfrac{1}{9} = (0.1121111111)^3 + \dfrac{1}{9} = 0.1125202246$

$x_3 = x_2^3 + \dfrac{1}{9} = (0.1125202246)^3 + \dfrac{1}{9} = 0.1125357073$

3.15 Increasing sequences

A sequence is increasing if each term is greater than the previous one.

Example

1 The terms of a sequence are given by u_n where u_n is given by the formula

$$u_n = \frac{n}{n+1}$$

(a) Write down the first three terms of this sequence.

(b) Prove, using proof by deduction, that the sequence generated is an increasing sequence.

Answer

1 (a) $u_1 = \dfrac{1}{1+1} = \dfrac{1}{2}$

$u_2 = \dfrac{2}{2+1} = \dfrac{2}{3}$

$u_3 = \dfrac{3}{3+1} = \dfrac{3}{4}$

First three terms are $\dfrac{1}{2}, \dfrac{2}{3}, \dfrac{3}{4}$

(b) If the first term is $\dfrac{n}{n+1}$ then the next term will be $\dfrac{n+1}{n+2}$.

If it is an increasing sequence then the difference between a term and the term before it will be positive.

Hence we can say $\dfrac{n+1}{n+2} - \dfrac{n}{n+1} > 0$

So, $\dfrac{(n+1)^2 - n(n+2)}{(n+2)(n+1)} > 0$

$\dfrac{n^2 + 2n + 1 - n^2 - 2n}{(n+2)(n+1)} > 0$

$\dfrac{1}{(n+2)(n+1)} > 0$

Now as n is a positive integer $(n+2)(n+1)$ will always be a positive number which means $\dfrac{1}{(n+2)(n+1)}$ will always be a positive fraction and therefore will always be greater than zero.

Hence it has been proved that the sequence is an increasing sequence.

> Remember that proof by deduction uses algebra as part of the proof.

3.16 Decreasing sequences

A sequence is decreasing if each term is less than the previous one.

3.17 Periodic sequences

A periodic sequence is a sequence that repeats itself after n terms, for example, the following is a periodic sequence:

$$1, 2, 3, 1, 2, 3, 1, 2, 3, \ldots$$

The *period* of this sequence is the number of terms in the repeating unit (the repeating unit in the sequence above is 1, 2, 3), so the period for this sequence is 3.

For example, the sequence generated by the relation $u_n = (-1)^n$ for $n > 0$ is as follows:

$$(-1)^1, (-1)^2, (-1)^3, (-1)^4, \ldots$$

which gives $-1, 1, -1, 1, \ldots$

This sequence is a periodic sequence with a period of 2.

Example

1 A sequence is generated using the following relation:

$$a_n = \frac{n+2}{2n-1}$$

State, showing your working, whether this sequence is an increasing, decreasing or periodic sequence.

. . . .

Answer

1 When $n = 1$, $a_1 = \dfrac{1+2}{2(1)-1} = 3$

$n = 3$, $a_3 = \dfrac{3+2}{2(3)-1} = 1$

$n = 50$, $a_{50} = \dfrac{50+2}{2(50)-1} = 0.5252$

$n = 1000$, $a_{1000} = \dfrac{1000+2}{2(1000)-1} = 0.5012$

The sequence is a decreasing sequence.

3.18 Using sequences and series in modelling

Sequences and series can be used to model situations in real life such as modelling the build-up of money when invested using compound interest or the way money is paid back for a loan. The following examples show some of these situations.

Examples

1 Aled decides to invest £1000 in a savings scheme on the first day of each year. The scheme pays 8% compound interest per annum, and interest is added on the last day of each year. The amount of savings, in pounds, at the end of the third year is given by:

$$1000 \times 1.08 + 1000 \times 1.08^2 + 1000 \times 1.08^3$$

Calculate, to the nearest pound, the amount of savings at the end of 30 years.

. . . .

Answer

1 After 30 years, the saving is represented by the following:

$$1000 \times 1.08 + 1000 \times 1.08^2 + \ldots + 1000 \times 1.08^{30}$$

This is a geometric progression with $a = 1000 \times 1.08$, $r = 1.08$ and $n = 30$

$$S_n = \frac{a(r^n - 1)}{r - 1} \quad \text{provided that } r \neq 1$$

$$S_{30} = \frac{1000 \times 1.08(1.08^{30} - 1)}{1.08 - 1} \approx £122\,346$$

2 The lengths of the sides of a fifteen-sided plane figure form an arithmetic sequence. The perimeter of the figure is 270 cm and the length of the largest side is eight times that of the smallest side. Find the length of the smallest side.

Answer

2 nth term, $t_n = a + (n - 1)d$

Hence 15th term $= a + 14d$

But the 15th term $= 8a$ (i.e. 8 times the smallest side which is the first term a)

Hence $a + 14d = 8a$, giving $a = 2d$

Now
$$S_n = \frac{n}{2}\big[2a + (n - 1)d\big]$$

$$S_{15} = \frac{15}{2}\big[2a + 14d\big]$$

$$270 = 15(a + 7d)$$

Since $a = 2d$ we can substitute this into the above equation to give

$$270 = 15(2d + 7d)$$

Solving gives $d = 2$

As $a = 2d$, $a = 4$

Hence, the smallest side = 4 cm

Test yourself

1. Expand $\dfrac{1 + x}{\sqrt{1 - 4x}}$ in ascending powers of x up to and including the term in x^2.
State the range of x for which the expansion is valid. [4]

2. Expand $(1 + 4x)^{\frac{1}{2}}$ in ascending powers of x as far as the term in x^2.
State the range of values of x for which your expansion is valid.
Expand $(1 + 4k + 16k^2)^{\frac{1}{2}}$ in ascending powers of k as far as the term in k^2. [6]

3. Expand $(1 + 2x)^{\frac{1}{2}}$ in ascending powers of x up to and including the term in x^3.
State the range of values of x for which the expansion is valid.
Expand $\sqrt{1.02}$ correct to six decimal places. [5]

4. Find an expression, in terms of n, for the sum of the first n terms of the arithmetic series: $4 + 10 + 16 + 22 + \ldots$.
Simplify your answer. [3]

5. The sum of the first seven terms of an arithmetic series is 182. The sum of the fifth and seventh terms of the series is 80. Find the first term and the common difference of the series. [4]

6. A geometric series has first term a and common ratio r. The sum of the first two terms of the geometric series is 2.7. The sum to infinity of the series is 3.6. Given that r is positive, find the values of r and a. [4]

7. In an arithmetic series the ninth term is double the fourth term. If the sixteenth term is 68, find the first term and the common difference of this arithmetic series. [5]

8. The nth term of a number sequence is denoted by t_n. The $(n + 1)$th term of the sequence satisfies: $t_{n+1} = 2t_n + 1$ for all positive integers n.
   (a) Given that $t_4 = 63$, evaluate t_1. [2]
   (b) Without carrying out any further calculations, explain why 6 043 582 cannot be one of the terms of this number sequence. [1]

9. Expand $6\sqrt{1 - 2x} - \dfrac{1}{1 + 4x}$

   in ascending powers of x up to and including the term in x^2. State the range of values of x for which your expansion is valid. [6]

Summary

Check you know the following facts:

The binomial expansion of (a + b)ⁿ for any rational value of n

$$(a + b)^n = a^n + \binom{n}{1}a^{n-1}b + \binom{n}{2}a^{n-2}b^2 + ... + \binom{n}{r}a^{n-r}b^r + ... + b^n$$

$$\binom{n}{r} = {}^nC_r = \frac{n!}{r!(n-r)!}$$

The binomial expansion of (1 + x)ⁿ for negative or fractional n

$$(1 + x)^n = 1 + nx + \frac{n(n-1)}{2!}x^2 + \frac{n(n-1)(n-2)}{3!}x^3 + ... \qquad |x| < 1$$

Sequences, arithmetic series and geometric series

The nth term of an arithmetic sequence

The nth term $t_n = a + (n-1)d$

where a is the first term, d is the common difference and n is the number of terms.

The sum to n terms of an arithmetic series

$$S_n = \frac{n}{2}\left[2a + (n-1)d\right]$$

The nth term of a geometric sequence

The nth term $t_n = ar^{n-1}$

where a is the first term, r is the common ratio and n is the number of terms.

The sum to n terms of a geometric sequence

$$S_n = \frac{a(1 - r^n)}{1 - r} \quad \text{provided } r \neq 1$$

The sum to infinity of a geometric sequence

$$S_\infty = \frac{a}{1 - r}$$

Note that for the sum to infinity to exist $|r| < 1$

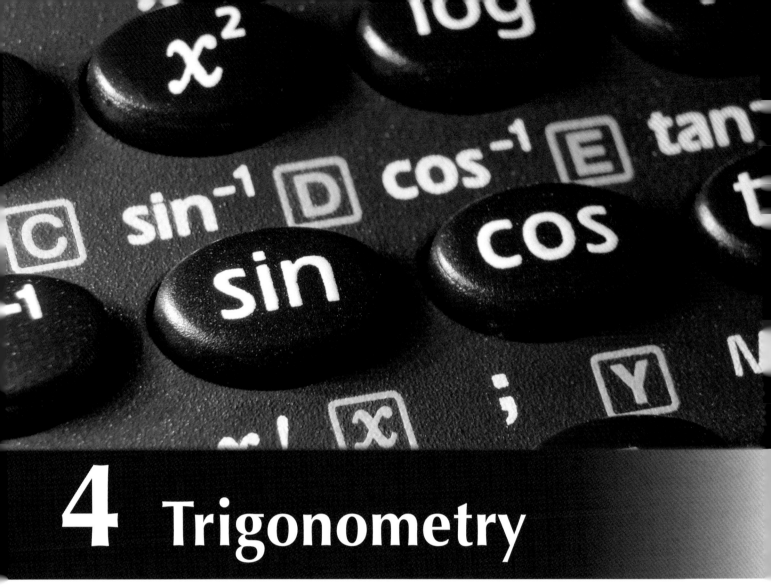

4 Trigonometry

Introduction

Trigonometry was covered in topic 5 of the AS and this topic builds on what you have already learned. Have a look back at topic 5 as a refresher before looking at this new material.

This topic covers the following:

4.1 Radian measure (arc length, area of sector and area of segment)

4.2 Use of small angle approximation for sine, cosine and tangent

4.3 Secant, cosecant and cotangent and their graphs

4.4 Inverse trigonometric functions $\sin^{-1}$, $\cos^{-1}$ and $\tan^{-1}$ and their graphs and domains

4.5 The trigonometric identities $\sec^2\theta = 1 + \tan^2\theta$ and $\csc^2\theta = 1 + \cot^2\theta$

4.6 Knowledge and use of the addition formulae $\sin(A \pm B)$, $\cos(A \pm B)$, $\tan(A \pm B)$ and geometric proofs of these addition formulae

4.7 Expressions for $a\cos\theta + b\sin\theta$ in the equivalent forms $R\cos(\theta \pm \alpha)$ or $R\sin(\theta \pm \alpha)$

4.8 Constructing proofs involving trigonometric functions and identities

4.1 Radian measure (arc length, area of sector and area of segment)

Radian measure was covered on page 103 of the AS book. Angles are usually measured in degrees but they can also be measured in radians. Here are some important facts you need to remember about radians:

$$1 \text{ radian} = \frac{180}{\pi} = 57.3°$$

π radians = 180° 2π radians = 360°

$\frac{\pi}{2}$ radians = 90° $\frac{\pi}{4}$ radians = 45°

$\frac{\pi}{3}$ radians = 60° $\frac{\pi}{6}$ radians = 30°

Arc length

> The length of an arc making an angle of θ radians at the centre $l = r\theta$

l is a fraction of the circumference and is given by

$$l = \frac{\theta}{2\pi} \times 2\pi r = r\theta$$

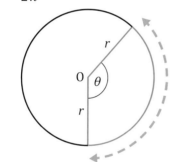

Remember that for arc lengths and areas of sectors the angle at the centre is measured in radians.

Area of a sector

> Area of a sector making an angle of θ radians at the centre $= \frac{1}{2}r^2\theta$

A is a fraction of the area of the complete circle and is given by

$$A = \frac{\theta}{2\pi} \times \pi r^2 = \frac{1}{2}r^2\theta$$

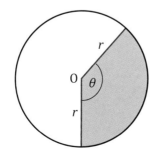

Area of a segment

A segment of a circle is the area bounded by a chord and an arc. It is the shaded area in the diagram below.

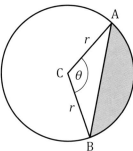

Area of sector ABC = $\frac{1}{2}r^2\theta$

Area of triangle ABC = $\frac{1}{2}ab \sin\theta$, but since $a = r$ and $b = r$ we can write

Area of triangle ABC = $\frac{1}{2}r^2 \sin\theta$

Area of segment = area of sector ABC − area of triangle ABC

$$= \frac{1}{2}r^2\theta - \frac{1}{2}r^2 \sin\theta$$

θ must be measured in radians.

> Area of segment = $\frac{1}{2}r^2(\theta - \sin\theta)$

Example

1

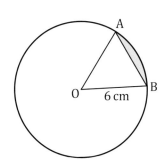

The diagram shows two points A and B on a circle with centre O and radius 6 cm. The length of the **arc** AB is 5.4 cm.

(a) Show that the area of the **sector** AOB is 16.2 cm²

(b) Find the area of the shaded region, giving your answer correct to one decimal place.

BOOST

Grade ⬆⬆⬆⬆

Check the formulae in the formula booklet regularly so you know which you need to remember. Note that the formulae for the length of an arc and the area of a sector are not given in the formula booklet.

· ·

Answer

1   (a)   Length of arc AB = $r\theta$

$$5.4 = 6\theta$$

$$\theta = 0.9 \text{ radians}$$

The angle at the centre is not given. The length of arc formula can be used to work out the angle at the centre in radians.

Area of sector AOB $= \frac{1}{2}r^2\theta$

$$= \frac{1}{2} \times 6^2 \times 0.9$$

$$= 16.2 \text{ cm}^2$$

(b) Area of triangle AOB $= \frac{1}{2}ab \sin C,$

$$= \frac{1}{2} \times 6 \times 6 \sin 0.9$$

$$= 14.10 \text{ cm}^2$$

> Note that both a and b are equal to the radius r of the circle.

Area of shaded region (i.e. the segment) = area of sector − area of triangle

$$= 16.2 - 14.10$$

$$= 2.1 \text{ cm}^2 \text{ (to 1 decimal place)}$$

> Remember to set your calculator to radian mode before working out this calculation.

Step by STEP

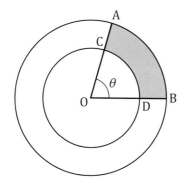

The diagram shows two concentric circles with common centre O. The radius of the larger circle is R cm and the radius of the smaller circle is r cm. The points A and B lie on the larger circle and are such that AÔB = θ radians. The smaller circle cuts OA and OB at points C and D respectively. The sum of the lengths of the arcs AB and CD is L cm. The area of the shaded region ACDB is K cm^2.

Given that AC = x cm find an expression for K in terms of x and L.

Steps to take

1   Find the lengths of the arcs AB and CD and then add them and equate them to L. This will then give L in terms of R, r and θ.

2   Find the areas of the sectors OAB and OCD. Subtract these two areas to give the shaded area. The resulting area will be equal to K and it will be in terms of R, r and θ.

3   Use the two equations, along with the fact that $R - r = x$ to find an expression for K in terms of x and L. This means that the variables θ and r will need to be eliminated.

Answer

Length of arc AB = $R\theta$ and length of arc CD = $r\theta$

$$L = R\theta + r\theta = \theta(R + r) \qquad (1)$$

$$\text{Area of sector OAB} = \frac{1}{2}R^2\theta$$

$$\text{Area of sector OCD} = \frac{1}{2}r^2\theta$$

Shaded area, $\qquad K = \frac{1}{2}R^2\theta - \frac{1}{2}r^2\theta = \frac{1}{2}\theta(R^2 - r^2) = \frac{1}{2}\theta(R + r)(R - r) \qquad (2)$

From equation (1) $\theta = \dfrac{L}{R + r}$ and substituting this in for θ in equation (2) we obtain:

$$K = \frac{1}{2}\left(\frac{L}{R + r}\right)(R + r)(R - r)$$

$$= \frac{1}{2}L(R - r)$$

Now, $\qquad\qquad R - r = x$

Hence, $\qquad\qquad K = \frac{1}{2}Lx$

> Notice that the $(R + r)$ in the numerator and the $(R + r)$ in the denominator can be cancelled.

Examples

1 The diagram shows two concentric circles with common centre O. The radius of the larger circle is R cm and the radius of the smaller circle is r cm. The points A and B lie on the larger circle and are such that $A\hat{O}B = \theta$ radians. The smaller circle cuts OA and OB at points C and D respectively. The length of the arc AB is L cm **greater** than the length of the arc CD. The area of the shaded region is K cm².

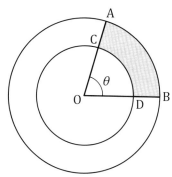

(a) (i)   Write down an expression for L in terms of R, r and θ.

(ii) Write down an expression for K in terms of R, r and θ.

(b) Use your results to part (a) to find an expression for r in terms of R, K and L.

> Remember that the length of an arc = $r\theta$, where r is the radius of the circle and θ is the angle at the centre measured in radians. This formula needs to be remembered.

Answer

1   (a)   (i)   arc AB − arc CD = L

$$R\theta - r\theta = L$$

Hence $\qquad L = \theta(R - r)$

(ii)   Shaded area = area of sector OAB – area of sector OCD

$$K = \frac{1}{2}R^2\theta - \frac{1}{2}r^2\theta$$

$$K = \frac{1}{2}\theta(R^2 - r^2)$$

(b) $K = \frac{1}{2}\theta(R^2 - r^2) = \frac{1}{2}\theta(R - r)(R + r)$

> $R^2 - r^2$ is the difference of two squares and can be factorised.

Now $L = \theta(R - r)$

Rearranging gives $\theta = \dfrac{L}{R - r}$

Substituting this into the equation for K gives

$$K = \frac{1}{2} \times \frac{L}{R - r} \times (R - r)(R + r)$$

Cancelling $(R - r)$ on the top and bottom gives

$$K = \frac{1}{2}L(R + r)$$

> When cancelling $(R - r)$ it is valid because
> $R - r \neq 0$, or $R \neq r$
> as otherwise there would be no shaded region.

$$\frac{2K}{L} = R + r$$

Giving $r = \dfrac{2K}{L} - R$

2

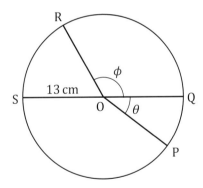

The diagram shows four points P, Q, R and S on a circle with centre O and radius 13 cm. The line QS is a diameter of the circle, PÔQ = θ radians and QÔR = ϕ radians.

(a)   The area of sector PÔQ is 60 cm².
Find the value of θ, giving your answer correct to two decimal places.

(b)   The length of the arc QR is 7 cm greater than the length of the arc RS.
Find the value of ϕ, giving your answer correct to two decimal places.

Answer

2   (a)   Area of sector POQ $= \frac{1}{2}r^2\theta$

$$60 = \frac{1}{2} \times 13^2\theta$$

$$60 = \frac{1}{2} \times 169\theta$$

$$\theta = \frac{120}{169} = 0.71 \text{ radians (to 2 decimal places)}$$

> Remember here that angles have to be in radians. Be careful to use π radians here rather than 180°.

     (b)   Length of arc QR $= 13\phi$

Length of arc RS $= r(\pi - \phi) = 13(\pi - \phi)$

$$QR - RS = 7$$

$$13\phi - 13(\pi - \phi) = 7$$

$$13\phi - 40.841 + 13\phi = 7$$

$$26\phi = 47.841$$

$$\phi = \frac{47.841}{26} = 1.84 \text{ radians (to 2 decimal places)}$$

4.2 Use of small angle approximation for sine, cosine and tangent

The small-angle approximation for sine, cosine and tangent is a useful simplification of the basic trigonometric functions which is approximately true in the limit where the angle approaches zero.

This means that if the angle is small and is measured in radians then the following approximations can be used:

$$\sin\theta \approx \theta$$
$$\cos\theta \approx 1 - \frac{\theta^2}{2}$$
$$\tan\theta \approx \theta$$

> All these apply only when angle θ is measured in radians.

Examples

1   Show that $\dfrac{\cos^2 x - 1}{x\sin 2x} \approx -\dfrac{1}{2}$ for small values of x measured in radians.

Answer

> Here we use a rearrangement of: $\sin^2 x + \cos^2 x = 1$
> to give: $\cos^2 x - 1 = -\sin^2 x$
> and also that:
> $\sin 2x = 2\sin x \cos x$.

1
$$\frac{\cos^2 x - 1}{x\sin 2x} = \frac{-\sin^2 x}{x\sin 2x}$$

$$= \frac{-\sin^2 x}{2x\sin x \cos x}$$

$$= \frac{-\sin x}{2x\cos x}$$

> $\dfrac{\sin x}{\cos x} = \tan x$

$$= -\frac{\tan x}{2x}$$

As x is a small angle measured in radians we have $\tan x \approx x$

Hence $\qquad \dfrac{\cos^2 x - 1}{x \sin 2x} \approx -\dfrac{x}{2x}$

$$\approx -\dfrac{1}{2}$$

---

2 Find a small positive value of x which is an approximate solution of the equation:

$$\cos x - 4 \sin x = x^2$$

. . . .

Answer

2 For small angles x we have $\sin x \approx x$ and $\cos x \approx 1 - \dfrac{x^2}{2}$

$$\cos x - 4 \sin x = x^2$$

$$1 - \dfrac{x^2}{2} - 4x \approx x^2$$

$$\dfrac{3}{2}x^2 + 4x - 1 \approx 0$$

$$3x^2 + 8x - 2 \approx 0$$

$$x = \dfrac{-b \pm \sqrt{b^2 - 4ac}}{2a} = \dfrac{-8 \pm \sqrt{64 + 24}}{6}$$

> This quadratic equation cannot be factorised so either the formula or completing the square can be used to find the solutions.

Hence $x = 0.230$ or -2.897

The only small positive value of x is $x = 0.230$ radians

The exact values of sin and cos for $0, \dfrac{\pi}{6}, \dfrac{\pi}{4}, \dfrac{\pi}{3}, \dfrac{\pi}{2}, \pi$ and multiples

The sine graph ($y = \sin \theta$) where θ is expressed in radians

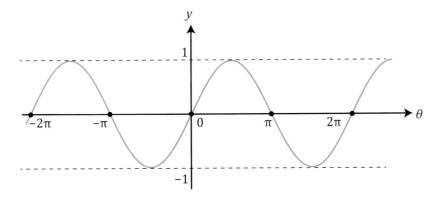

> The sine graph has a period of 2π meaning the graph repeats itself every 2π radians.

The sine graph has a period of 2π as a particular value of θ will have the same y-value at an angle of $\theta + 2\pi$, $\theta + 4\pi$, and so on.

The cosine graph ($y = \cos\theta$) where θ is expressed in radians

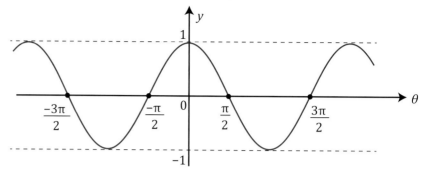

The cosine graph has a period of 2π meaning the graph repeats itself every 2π radians.

The cosine graph has a period of 2π as a particular value of θ will have the same y-value at an angle of $\theta + 2\pi$, $\theta + 4\pi$, and so on.

The tangent graph ($y = \tan\theta$) where θ is expressed in radians

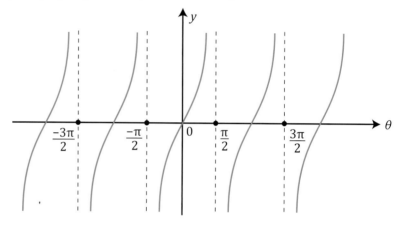

The period of the $\tan\theta$ graph is π radians.

The exact values of the sine, cosine and tangent of $\frac{\pi}{4}$

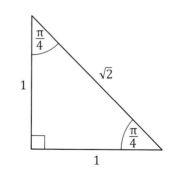

$$\sin\frac{\pi}{4} = \frac{\text{opposite}}{\text{hypotenuse}} = \frac{1}{\sqrt{2}}$$

$$\cos\frac{\pi}{4} = \frac{\text{adjacent}}{\text{hypotenuse}} = \frac{1}{\sqrt{2}}$$

$$\tan\frac{\pi}{4} = \frac{\text{opposite}}{\text{adjacent}} = \frac{1}{1} = 1$$

The exact values of the sine, cosine and tangent of $\frac{\pi}{6}$ and $\frac{\pi}{3}$

The exact values can be worked out using an equilateral triangle having sides of length 2 and then drawing in one of the lines of symmetry to form two identical right-angled triangles:

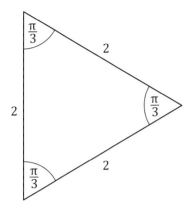

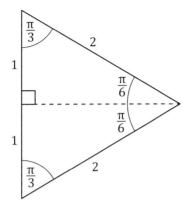

Start off with an equilateral triangle having length of side = 2. All the angles of the triangle will be $\frac{\pi}{3}$.

The perpendicular bisector divides the original triangle into two right-angled triangles. Notice that the angle is bisected as well as a side.

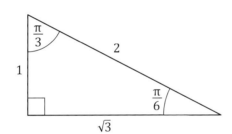

Half of the original triangle is used. The length of the base of this triangle is worked out using Pythagoras' theorem.

Base = $\sqrt{2^2 - 1^2} = \sqrt{3}$

$$\sin\frac{\pi}{6} = \frac{1}{2} \qquad \sin\frac{\pi}{3} = \frac{\sqrt{3}}{2}$$

$$\cos\frac{\pi}{6} = \frac{\sqrt{3}}{2} \qquad \cos\frac{\pi}{3} = \frac{1}{2}$$

$$\tan\frac{\pi}{6} = \frac{1}{\sqrt{3}} \qquad \tan\frac{\pi}{3} = \sqrt{3}$$

If you are asked to find the exact value you must not approximate any of the sides as a decimal.

4.3 Secant, cosecant and cotangent and their graphs

Sec θ

Sec θ is the reciprocal of $\cos\theta$ so $\boxed{\sec\theta = \dfrac{1}{\cos\theta}}$

The graph of $y = \sec\theta$ is shown on the following page.

You can see from the graph below that the curve $y = \sec\theta$ is defined for all values of θ other than $\theta \pm\frac{\pi}{2}$, $\pm\frac{3\pi}{2}$ etc., since, at these values of θ, $\cos\theta = 0$. Notice that as θ approaches these values, the value of y approaches $\pm\infty$ (infinity) and the vertical lines, $\theta = \pm\frac{\pi}{2}$, $\pm\frac{3\pi}{2}$ etc., are called asymptotes to the curve.

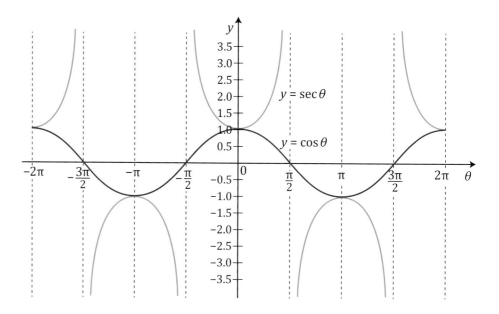

Cosec θ

Cosec θ is the reciprocal of $\sin\theta$ so

$$\csc\theta = \frac{1}{\sin\theta}$$

The graph of $y = \csc\theta$ is shown below. The curve $y = \csc\theta$ is defined for all values of θ other than where $\sin\theta = 0$, i.e. $\theta = 0$, $\pm\pi$, $\pm 2\pi$, etc., which are asymptotes to the curve.

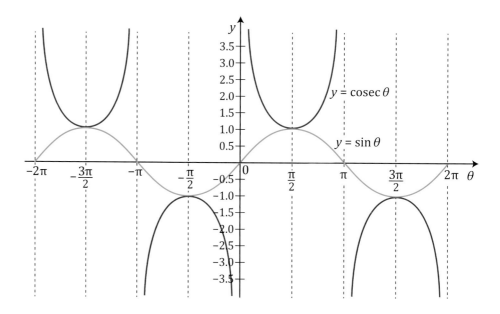

Cot θ

Cot θ is the reciprocal of tan θ so $\boxed{\cot \theta = \dfrac{1}{\tan \theta}}$

The graph of $y = \cot \theta$ is shown below. The curve $y = \cot \theta$ is defined for all values of θ other than where $\tan \theta = 0$, i.e. $\theta = 0$, $\pm\pi$, $\pm2\pi$, etc., which are asymptotes to the curve.

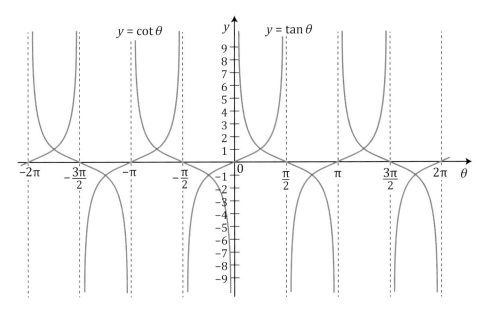

4.4 Inverse trigonometric functions sin⁻¹, cos⁻¹ and tan⁻¹ and their graphs and domains

The graph of $y = \sin^{-1} \theta$

To obtain the graph of $y = \sin^{-1} \theta$, first take the graph of $y = \sin \theta$ in the region between $\theta = -\frac{\pi}{2}$ and $\theta = \frac{\pi}{2}$ and reflect it in the line $y = \theta$.

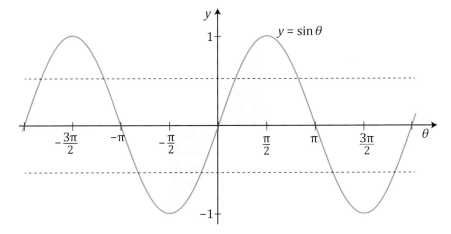

If you look at the curve for $y = \sin \theta$, one value of y corresponds to many values of θ. However, to find an inverse it is necessary that one value of y corresponds to only one value of θ, and this is known as **one-to-one**.

For this reason, the values of θ are restricted to the interval between $-\frac{\pi}{2}$ and $\frac{\pi}{2}$

The graph for $y = \sin^{-1}\theta$ is obtained by reflecting the graph of $y = \sin\theta$ in the line $y = \theta$.

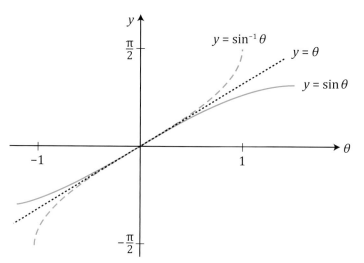

Notice from the graph that, for $y = \sin^{-1}\theta$, the only allowable values of θ lie in the interval $[-1, 1]$, i.e. $-1 \le \theta \le 1$. The set of allowable values that can be entered into a function is called the domain.

The range is the corresponding set of y-values the function can have.

The range here is $-\frac{\pi}{2} \le \theta \le \frac{\pi}{2}$.

The graph of $y = \cos^{-1}\theta$

To obtain the graph of $y = \cos^{-1}\theta$, first take the graph of $y = \cos\theta$ in the region between $\theta = 0$ and $\theta = \pi$ and reflect it in the line $y = \theta$.

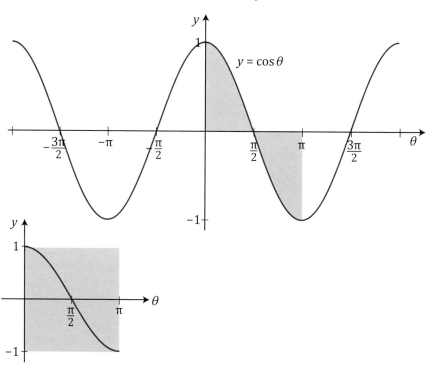

So that the original function is one-to-one, the domain is restricted to $0 \leq \theta \leq \pi$.

When the function $(y = \cos\theta)$ is reflected in the line $y = \theta$, the graph of the inverse function $y = \cos^{-1}\theta$ is obtained as shown here:

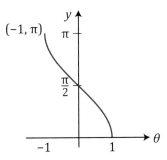

The domain of the original function (i.e. $0 \leq \theta \leq \pi$) becomes the range of the inverse function and the range of the original function (i.e. $-1 \leq \cos\theta \leq 1$) will become the domain of the inverse function.

The domain of $y = \cos^{-1}\theta$ is $[-1, 1]$, i.e. $-1 \leq \cos\theta \leq 1$.

The range of $y = \cos^{-1}\theta$ is $[0, \pi]$, i.e. $0 \leq \cos^{-1}\theta \leq \pi$.

The graph of $y = \tan^{-1}\theta$

Here the graph of $y = \tan\theta$ is reflected in the line $y = \theta$ to produce the graph of the inverse function $y = \tan^{-1}\theta$.

Again, only part of the graph is used so that each y-value has only one possible x-value.

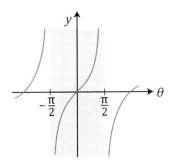

The graph of $y = \tan^{-1}\theta$ is shown below.

The domain of $y = \tan^{-1}\theta$ is the set of all real numbers

The range of $y = \tan^{-1}\theta$ is $\left(-\frac{\pi}{2}, \frac{\pi}{2}\right)$, i.e. $-\frac{\pi}{2} < \tan^{-1}\theta < \frac{\pi}{2}$.

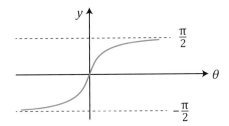

You may see the identity sign ≡ being used in place of the equals sign.

4.5 The trigonometric identities $\sec^2\theta = 1 + \tan^2\theta$ and $\operatorname{cosec}^2\theta = 1 + \cot^2\theta$

Here are two more trigonometric identities you need to remember and use:

$$\sec^2\theta = 1 + \tan^2\theta$$

$$\operatorname{cosec}^2\theta = 1 + \cot^2\theta$$

Both of these formulae must be remembered as they are not included in the formula booklet.

Proof of the two trigonometric identities, $\sec^2\theta = 1 + \tan^2\theta$ and $\operatorname{cosec}^2\theta = 1 + \cot^2\theta$

These two trigonometric identities may be proved in the following ways:

Using $\tan\theta = \dfrac{\sin\theta}{\cos\theta}$ and $\sin^2\theta + \cos^2\theta = 1$

Dividing through by $\cos^2\theta$ gives:

$$\frac{\sin^2\theta}{\cos^2\theta} + 1 = \frac{1}{\cos^2\theta}$$

$$\left(\frac{\sin\theta}{\cos\theta}\right)^2 + 1 = \left(\frac{1}{\cos\theta}\right)^2$$

$$\tan^2\theta + 1 = \sec^2\theta$$

Similarly, dividing through by $\sin^2\theta$ gives:

$$\frac{\cos^2\theta}{\sin^2\theta} + 1 = \frac{1}{\sin^2\theta}$$

$$\left(\frac{\cos\theta}{\sin\theta}\right)^2 + 1 = \left(\frac{1}{\sin\theta}\right)^2$$

$$\cot^2\theta + 1 = \operatorname{cosec}^2\theta$$

Solution of trigonometric equations making use of the identities $\sec^2\theta = 1 + \tan^2\theta$ and $\operatorname{cosec}^2\theta = 1 + \cot^2\theta$

You will frequently be asked to solve equations where you have to make use of the trigonometric identities $\sec^2\theta = 1 + \tan^2\theta$ and $\operatorname{cosec}^2\theta = 1 + \cot^2\theta$. In many of these questions, you need to form a quadratic equation in terms of just one trigonometric function before solving it, using factorisation where possible.

The following examples will explain this technique.

Examples

In this type of question, retain the term which is to the power 1 (i.e. $\tan\theta$).

1   Find the values of θ in the range $0° \le \theta \le 360°$ that satisfy the equation

$$\sec^2\theta + 5 = 5\tan\theta$$

giving your answers to one decimal place.

Answer

1

$$\sec^2 \theta + 5 = 5 \tan \theta$$

$$(1 + \tan^2 \theta) + 5 = 5 \tan \theta$$

$$\tan^2 \theta - 5 \tan \theta + 6 = 0$$

$$(\tan \theta - 3)(\tan \theta - 2) = 0$$

A quadratic equation in terms of $\tan \theta$ is formed which is then factorised and solved to determine the values of θ in the range specified in the question.

Solving gives $\tan \theta = 3$ or $\tan \theta = 2$

$$\theta = \tan^{-1}(3)$$

giving $\theta = 71.6°$ or $251.6°$ (correct to one decimal place)

or $\qquad \theta = \tan^{-1}(2)$

giving $\theta = 63.4°$ or $243.4°$ (correct to one decimal place)

$\theta = 63.4°, 71.6°, 243.4°$ or $251.6°$ (all correct to one decimal place)

Always list all your solutions in numerical order as a final answer.

Use $\sec^2 \theta = 1 + \tan^2 \theta$ to give the equation just in terms of $\tan \theta$.

The graph of $y = \tan \theta$ has a period of 180°, so once a solution is found another solution may be found by adding 180°.

Another way to find the values of θ is to use the CAST method. Tan θ is positive in the first and third quadrants.

---

2 Find the values of θ in the range $0° \leq \theta \leq 360°$ that satisfy the equation

$$2 \tan^2 \theta = 6 \sec \theta - 6$$

Answer

2

$$2(\sec^2 \theta - 1) = 6 \sec \theta - 6$$

$$2 \sec^2 \theta - 2 = 6 \sec \theta - 6$$

$$2 \sec^2 \theta - 6 \sec \theta + 4 = 0$$

$$\sec^2 \theta - 3 \sec \theta + 2 = 0$$

$$(\sec \theta - 2)(\sec \theta - 1) = 0$$

$$\sec \theta = 2 \quad \text{or} \quad \sec \theta = 1$$

$$\frac{1}{\cos \theta} = 2 \quad \text{or} \quad \frac{1}{\cos \theta} = 1$$

$$\cos \theta = \frac{1}{2} \quad \text{giving } \theta = 60° \text{ or } 300°$$

$$\cos \theta = 1 \quad \text{giving } \theta = 0° \text{ or } 360°$$

Hence $\theta = 0°, 60°, 300°$ or $360°$

Retain the term which is to the power 1 (i.e. $\sec \theta$).

Use the formula $\sec \theta = \dfrac{1}{\cos \theta}$
You must remember this.

Use a graph of $y = \cos \theta$ or the CAST method to find the angles in the required range.

Using the CAST method $\cos \theta$ is positive in the 1st and 4th quadrants.

As $\qquad \cos^{-1}\left(\dfrac{1}{2}\right) = 60°$

and this is in the 1st quadrant, the other angle in the range will be $360° - 60° = 300°$.

The angles which give $\cos \theta = 1$, can be worked out in a similar way.

4.6 Knowledge and use of the addition formulae $\sin(A \pm B)$, $\cos(A \pm B)$ and $\tan(A \pm B)$

There are a number of trigonometric identities which are used when solving trigonometric equations or integrating a trigonometric function.

Trigonometric identities

These are all given in the formula booklet and can be looked up.

$$\sin(A \pm B) = \sin A \cos B \pm \cos A \sin B$$

$$\cos(A \pm B) = \cos A \cos B \mp \sin A \sin B$$

$$\tan(A \pm B) = \frac{\tan A \pm \tan B}{1 \mp \tan A \tan B}$$

Double angle formulae

$$\sin 2A = 2\sin A \cos A$$

$$\cos 2A = \cos^2 A - \sin^2 A$$
$$= 1 - 2\sin^2 A$$
$$= 2\cos^2 A - 1$$

$$\tan 2A = \frac{2\tan A}{1 - \tan^2 A}$$

Important rearrangements

Here are some important rearrangements of the above formulae. These are useful for proving some identities and also when integrating expressions.

These rearrangements are obtained by combining the double angle formulae with the identity
$$\sin^2 A + \cos^2 A = 1$$

$$\sin^2 A = \frac{1}{2}\left(1 - \cos 2A\right)$$

$$\cos^2 A = \frac{1}{2}\left(1 + \cos 2A\right)$$

Here is an example of how one of these rearrangements can be used:

Suppose you have to find $\int(2 + \cos^2 \theta)\mathrm{d}\theta$

Change from
$$\cos^2 \theta \text{ to } \frac{1 + \cos 2\theta}{2}$$
because it is easier to integrate $\cos 2\theta$ than $\cos^2 \theta$.

$$\int(2 + \cos^2 \theta)\mathrm{d}\theta = \int\left(2 + \frac{1}{2}\left(1 + \cos 2\theta\right)\right)\mathrm{d}\theta$$

$$= \frac{1}{2}\int\left(5 + \cos 2\theta\right)\mathrm{d}\theta$$

$$= \frac{1}{2}\left(5\theta + \frac{1}{2}\sin 2\theta\right) + c$$

4.6 Knowledge and use of the addition formulae sin (A ± B), cos (A ± B) and tan (A ± B)

Examples

1 Showing all your working, find the values of θ in the range $0° \leq \theta \leq 360°$ satisfying the equation $\cos 2\theta = \sin \theta$.

Change $\cos 2\theta$ to $1 - 2\sin^2 \theta$ so that the equation becomes an equation just involving sin.

Answer

1
$$1 - 2\sin^2 \theta = \sin \theta$$

$$2\sin^2 \theta + \sin \theta - 1 = 0$$

$$(2\sin \theta - 1)(\sin \theta + 1) = 0$$

Notice that this is a quadratic equation in $\sin \theta$ that can be factorised and hence solved.

Hence $\sin \theta = \dfrac{1}{2}, -1$

When $\sin \theta = \dfrac{1}{2}$, $\theta = 30°, 150°$

Use the graphical or CAST method to find the angles in the required range.

When $\sin \theta = -1$, $\theta = 270°$

BOOST

Grade ⬆⬆⬆⬆

Always check that you only include the solutions in the range specified in the question.

2 Given that $2\cos 2\theta + 3\sin \theta = 3$, show that

$$4\sin^2 \theta - 3\sin \theta + 1 = 0$$

Notice the double angle here. Use $\cos 2\theta = 1 - 2\sin^2 \theta$. The reason this identity is used rather than one of the others for $\cos 2\theta$ is that we can create a quadratic equation in sin only.

Answer

2
$$2\cos 2\theta + 3\sin \theta = 3$$

$$2(1 - 2\sin^2 \theta) + 3\sin \theta = 3$$

$$2 - 4\sin^2 \theta + 3\sin \theta = 3$$

$$4\sin^2 \theta - 3\sin \theta + 1 = 0$$

3 (a) Prove the identity $\cos 3\theta = 4\cos^3 \theta - 3\cos \theta$.

(b) Solve the equation

$$\cos 3\theta + \cos^2 \theta = 0$$

finding the values of θ in the range $0° \leq \theta \leq 360°$.

Answer

3 (a) $\cos 3\theta = \cos (2\theta + \theta)$

Here the identity
$\cos (A + B) = \cos A \cos B - \sin A \sin B$
is used.

$$= \cos 2\theta \cos \theta - \sin 2\theta \sin \theta$$

$$= (2\cos^2 \theta - 1) \cos \theta - 2\sin \theta \cos \theta \sin \theta$$

$$= (2\cos^2 \theta - 1) \cos \theta - 2\sin^2 \theta \cos \theta$$

$$= (2\cos^2 \theta - 1) \cos \theta - 2(1 - \cos^2 \theta)\cos \theta$$

$$= 4\cos^3 \theta - 3\cos \theta$$

(b)
$$\cos 3\theta + \cos^2 \theta = 0$$

$$4\cos^3 \theta - 3\cos \theta + \cos^2 \theta = 0$$

$$\cos \theta(4\cos^2 \theta + \cos \theta - 3) = 0$$

$$\cos \theta = 0 \quad \text{or} \quad 4\cos^2 \theta + \cos \theta - 3 = 0$$

$$\cos\theta = 0 \quad \text{or} \quad (4\cos\theta - 3)(\cos\theta + 1) = 0$$

$$\cos\theta = 0 \quad \text{or} \quad \cos\theta = \frac{3}{4} \quad \text{or} \quad \cos\theta = -1$$

$$\theta = 41.4°, 90°, 180°, 270°, 318.6°$$

4　Prove the identity $\dfrac{1 - \cos 2\theta}{\sin 2\theta} = \tan\theta$

. . . .

Answer

Use the double angle formula to remove the double angles.

4　　$\dfrac{1 - \cos 2\theta}{\sin 2\theta} = \dfrac{1 - \cos 2\theta}{2\sin\theta\cos\theta}$

$$= \dfrac{1 - (1 - 2\sin^2\theta)}{2\sin\theta\cos\theta}$$

Divide the numerator and denominator by $2\sin\theta$.

$$= \dfrac{2\sin^2\theta}{2\sin\theta\cos\theta}$$

$$= \dfrac{\sin\theta}{\cos\theta}$$

$$= \tan\theta$$

5　Find $\int\cos^2 x \, dx$

. . . .

Answer

Express $\cos^2 x$ in terms of the double angle $\cos 2x$ using a double angle formula.

5　　$\displaystyle\int\cos^2 x \, dx = \int\dfrac{\cos 2x + 1}{2} \, dx$

$$= \dfrac{1}{2}\int\left(\cos 2x + 1\right) dx$$

It is usually best to take any constant terms outside the integral sign before integrating as it makes the integration easier.

$$= \dfrac{1}{2}\left(\dfrac{1}{2}\sin 2x + x\right) + c$$

$$= \dfrac{1}{4}\sin 2x + \dfrac{x}{2} + c$$

6　Show that $\displaystyle\int_0^{\frac{\pi}{2}}\sin^2\theta \, d\theta = \dfrac{\pi}{4}$

. . . .

Answer

6　　$\displaystyle\int_0^{\frac{\pi}{2}}\sin^2\theta \, d\theta = \int_0^{\frac{\pi}{2}}\dfrac{1 - \cos 2\theta}{2} \, d\theta$

$$= \dfrac{1}{2}\int_0^{\frac{\pi}{2}}\left(1 - \cos 2\theta\right) d\theta$$

$$= \dfrac{1}{2}\left[\theta - \dfrac{\sin 2\theta}{2}\right]_0^{\frac{\pi}{2}}$$

$$= \frac{1}{2}\left[\left(\frac{\pi}{2} - \frac{1}{2}\sin\pi\right) - \left(0 - \frac{1}{2}\sin 0\right)\right]_0^{\frac{\pi}{2}}$$

$$= \frac{1}{2}\left[\left(\frac{\pi}{2} - 0\right) - \left(0 - 0\right)\right]$$

$$= \frac{\pi}{4}$$

---

7 Showing all your working, find the values of θ in the range $0° \leq \theta \leq 360°$ satisfying the equation $\sin 2\theta = \sin \theta$.

. . . .

Answer

7 $$\sin 2\theta = \sin \theta$$

$$2 \sin \theta \cos \theta = \sin \theta$$

$$2 \sin \theta \cos \theta - \sin \theta = 0$$

$$\sin \theta (2\cos \theta - 1) = 0$$

Hence either $\sin \theta = 0$ or $\cos \theta = \frac{1}{2}$

When $\sin \theta = 0$, $\theta = 0°, 180°, 360°$

When $\cos \theta = \frac{1}{2}$, $\theta = 60°, 300°$

Hence $\theta = 0°, 60°, 180°, 300°, 360°$

> Use the double angle formula.
> $\sin 2A = 2 \sin A \cos A$.

> Do not divide through by $\sin \theta$ or you will lose one of the solutions. Instead take $\sin \theta$ out of the brackets as a factor.

---

8 (a) Show that $3 \sin \theta - \cos 2\theta \equiv 2 \sin^2 \theta + 3 \sin \theta - 1$ for all values of θ.

(b) Find the values of θ in the range $0° \leq \theta \leq 360°$ satisfying the equation $3 \sin \theta - \cos 2\theta + 2 = 0$.

. . . .

Answer

8 (a) $3 \sin \theta - \cos 2\theta = 3 \sin \theta - (\cos^2 \theta - \sin^2 \theta)$

$$= 3 \sin \theta - \cos^2 \theta + \sin^2 \theta$$

$$= 3 \sin \theta - (1 - \sin^2 \theta) + \sin^2 \theta \qquad \text{Use } \cos^2 \theta = 1 - \sin^2 \theta$$

$$= 2 \sin^2 \theta + 3 \sin \theta - 1$$

(b) $3 \sin \theta - \cos 2\theta \equiv 2 \sin^2 \theta + 3 \sin \theta - 1$

$$3 \sin \theta - \cos 2\theta + 2 = 2 \sin^2 \theta + 3 \sin \theta - 1 + 2$$

$$0 = 2 \sin^2 \theta + 3 \sin \theta + 1$$

$$(2 \sin \theta + 1)(\sin \theta + 1) = 0$$

$$\sin \theta = -\frac{1}{2}, -1$$

$$\theta = 210°, 270°, 330°$$

> Use the double angle formula.
> $\cos 2A = \cos^2 A - \sin^2 A$.
> Note that this is not given directly in the formula booklet.

Geometric proofs of the addition formulae

You are required to understand the proofs for the following addition formulae:

$$\sin (A \pm B) = \sin A \cos B \pm \cos A \sin B$$

$$\cos (A \pm B) = \cos A \cos B \mp \sin A \sin B$$

$$\tan (A \pm B) = \frac{\tan A \pm \tan B}{1 \mp \tan A \tan B}$$

Here we will just prove the formulae for $\sin (A + B)$.

We first start off with the following diagram.

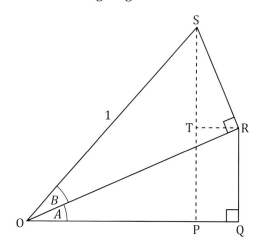

In the diagram length OS = 1 and notice that triangles OSR and ORQ are right-angled triangles.

Notice that if we want to find $\sin (A + B)$ we need to find PS and to find PS we will need to find PT and TS.

In triangle OSR, $\dfrac{\text{OR}}{1} = \cos B$ so OR = $\cos B$

In triangle ORQ, $\dfrac{\text{QR}}{\text{OR}} = \sin A$ so QR = OR $\sin A$ = $\cos B \sin A$

Now   PT = QR so PT = $\cos B \sin A$

In triangle TSR, angle TRO = angle A (alternate angles) and angle TRS = 90 – A.

Hence   angle RST = angle A

$$\frac{\text{RS}}{1} = \sin B \text{ so RS} = \sin B$$

$$\frac{\text{TS}}{\text{RS}} = \cos \text{RST, however } \cos \text{RST} = \cos A$$

Hence   TS = RS $\cos A$ = $\sin B \cos A$

Now   PS = PT + TS = $\cos B \sin A + \sin B \cos A$

$$\sin (A + B) = \frac{\text{PS}}{1} = \sin A \cos B + \cos A \sin B$$

Active Learning

See if you can work out the proofs for the other addition formulae. If you get stuck look at some YouTube videos or do some research using the Internet

Make sure you cover the proofs of all the formulae.

4.7 Expressions for $a \cos \theta + b \sin \theta$ in the equivalent forms, $R \cos(\theta \pm \alpha)$ or $R \sin(\theta \pm \alpha)$

Expressions in the form $a \cos \theta + b \sin \theta$ can be expressed in the alternative forms $R \cos(\theta \pm \alpha)$ or $R \sin(\theta \pm \alpha)$.

To express $4 \cos \theta + 2 \sin \theta$ in the form $R \cos(\theta - \alpha)$ take the following steps.

$$4 \cos \theta + 2 \sin \theta \equiv R \cos(\theta - \alpha)$$

$$4 \cos \theta + 2 \sin \theta \equiv R \cos \theta \cos \alpha + R \sin \theta \sin \alpha$$

> Use the trig identity
> $\cos(A - B) = \cos A \cos B + \sin A \sin B$

As the coefficients of $\cos \theta$ must be the same on both sides, so

$$R \cos \alpha = 4$$

Similarly the coefficients of $\sin \theta$ are the same on both sides, so

$$R \sin \alpha = 2$$

> The relationship is an identity, i.e. true for all values of θ. Hence we use $\equiv$ instead of $=$.

Dividing these two equations gives $\dfrac{R \sin \alpha}{R \cos \alpha} = \tan \alpha = \dfrac{1}{2}$

$$\alpha = \tan^{-1} \frac{1}{2} = 26.6°$$

$$R^2 \sin^2 \alpha + R^2 \cos^2 \alpha = 2^2 + 4^2$$

$$R^2(\sin^2 \alpha + \cos^2 \alpha) = 2^2 + 4^2$$

$$R^2 = 2^2 + 4^2$$

$$R = \sqrt{2^2 + 4^2} = \sqrt{20}$$

> Remember that
> $\sin^2 \alpha + \cos^2 \alpha = 1$

The two values R and α have been found so

$$4 \cos \theta + 2 \sin \theta = \sqrt{20} \cos(\theta - 26.6°)$$

Finding the greatest and least values of a trigonometric function

The greatest or least value of $\sin \theta$ or $\cos \theta$ is 1 or -1 respectively.

For the expression $5 \cos \theta$ the greatest and least values would be 5 and -5 respectively.

For the expression $5 \cos(\theta - 30°)$ the greatest and least values would be 5 and -5 respectively.

Examples

1 Find the greatest and least values of $\dfrac{1}{6 + 5 \sin(x + 30°)}$

· ·

Answer

1   $5 \sin(x + 30°)$ has greatest and least values of 5 and -5.

Least value of $\dfrac{1}{6 + 5 \sin(x + 30°)}$ would be $\dfrac{1}{6 + 5} = \dfrac{1}{11}$

Greatest value of $\dfrac{1}{6 + 5 \sin(x + 30°)}$ would be $\dfrac{1}{6 - 5} = 1$

> Notice the least value occurs when the denominator is largest and the greatest value occurs when the denominator is least.

2 Given that 5 cos x + 12 sin x = 13 cos (x – 67.4°), find the greatest and least values of 5 cos x + 12 sin x and write down a value for x for which the least value occurs.

Answer

2   5 cos x + 12 sin x = 13 cos (x – 67.4°)

The greatest and least values of cos (x – 67.4°) are 1 and –1.

Hence   greatest value of 5 cos x + 12 sin x = 13 × 1 = 13

         least value of 5 cos x + 12 sin x = 13 × –1 = –13

Least value occurs when cos (x – 67.4°) = –1

Now cos 180 = –1

Hence x – 67.4° = 180°

So x = 247.4°

Other possible values are 540°, 900°, etc.

3 If $\sqrt{3}$ cos θ + sin θ = 2 cos (θ – 30°), find the greatest and least values of

$$\frac{1}{\sqrt{3} \cos \theta + \sin \theta - 3}$$

Write down a value of θ for which the greatest value occurs.

Answer

3   $\dfrac{1}{\sqrt{3} \cos \theta + \sin \theta - 3} = \dfrac{1}{2 \cos (\theta - 30°) - 3}$

The maximum and minimum values of cos (θ – 30°) are 1 and –1 respectively.

Hence   the minimum value of $\dfrac{1}{\sqrt{3} \cos \theta + \sin \theta - 3}$ is $\dfrac{1}{2 - 3} = -1$

    the maximum value of $\dfrac{1}{\sqrt{3} \cos \theta + \sin \theta - 3}$ is $\dfrac{1}{-2 - 3} = -\dfrac{1}{5}$

For the maximum value of $\dfrac{1}{\sqrt{3} \cos \theta + \sin \theta - 3}$

where   cos (θ – 30°) = –1

so θ – 30° = 180°

and θ = 210°

4 If sin θ + $\sqrt{3}$ cos θ = 2sin (θ + 60°), find the greatest and least values of

$$\frac{1}{\sin \theta + \sqrt{3} \cos \theta + 5}$$

. . . .

Answer

4   $\dfrac{1}{\sin \theta + \sqrt{3} \cos \theta + 5} = \dfrac{1}{2 \sin(\theta + 60°) + 5}$

The maximum and minimum values of $\cos(\theta + 60°)$ are 1 and −1 respectively.

Hence   the minimum value of $\dfrac{1}{\sin \theta + \sqrt{3} \cos \theta + 5}$ is $\dfrac{1}{2 + 5} = \dfrac{1}{7}$

the maximum value of $\dfrac{1}{\sin \theta + \sqrt{3} \cos \theta + 5}$ is $\dfrac{1}{-2 + 5} = \dfrac{1}{3}$

---

5 (a) Express $3 \cos \theta + 4 \sin \theta$ in the form $R \cos(\theta - \alpha)$, where R and α are constants with $R > 0$ and $0° \le \alpha \le 90°$.

(b) Use your results to part (a) to find the least value of

$$\dfrac{1}{3 \cos \theta + 4 \sin \theta + 10}$$

Write down a value for θ for which this least value occurs.

. . . .

Answer

5 (a) $3 \cos \theta + 4 \sin \theta = R \cos(\theta - \alpha)$

$= R \cos \theta \cos \alpha + R \sin \theta \sin \alpha$

$R \cos \alpha = 3$ and $R \sin \alpha = 4$

$\dfrac{R \sin \alpha}{R \cos \alpha} = \tan \alpha = \dfrac{4}{3}$

$\tan \alpha = \dfrac{4}{3}$ so $\alpha = 53.1°$

$R = \sqrt{3^2 + 4^2} = \sqrt{25} = 5$

Only the positive value for R is used here because $R > 0$ is stated in the question.

Hence $3 \cos \theta + 4 \sin \theta = 5 \cos(\theta - 53.1°)$

(b) $\dfrac{1}{3 \cos \theta + 4 \sin \theta + 10} = \dfrac{1}{5 \cos(\theta - 53.1°) + 10}$

The least value occurs when the denominator is greatest. The greatest value of the cos function is +1.

Hence least value is $\dfrac{1}{5 + 10} = \dfrac{1}{15}$

This value occurs when $\cos(\theta - 53.1°) = 1$

$\cos^{-1} 1 = 0°$

Hence $\theta - 53.1° = 0$

So $\theta = 53.1°$

---

6 (a) Express $\cos \theta + \sqrt{3} \sin \theta$ in the form $R \sin(\theta + \alpha)$, where $R > 0$ and $0° < \alpha < 90°$.

(b) Find all the values of θ in the range $0° < \theta < 360°$ satisfying the equation

$$\cos \theta + \sqrt{3} \sin \theta = 1$$

Answer

6   (a) $\cos\theta + \sqrt{3}\sin\theta \equiv R\sin(\theta + \alpha)$

$\cos\theta + \sqrt{3}\sin\theta \equiv R\sin\theta\cos\alpha + R\cos\theta\sin\alpha$

$R\sin\alpha = 1,\ R\cos\alpha = \sqrt{3}$

$\therefore \tan\alpha = \dfrac{1}{\sqrt{3}}$ so $\alpha = 30°$

$R = \sqrt{1 + 3} = 2$

Hence, $\cos\theta + \sqrt{3}\sin\theta \equiv 2\sin(\theta + 30°)$

(b) $\cos\theta + \sqrt{3}\sin\theta = 1$

$2\sin(\theta + 30°) = 1$

$\sin(\theta + 30°) = \dfrac{1}{2}$

$\theta + 30° = 30°,\ 150°,\ 390°$

$\theta = 0°,\ 120°,\ 360°$

4.8 Constructing proofs involving trigonometric functions and identities

Showing by counter example

To prove that a given statement is false, you need to find *just one case* for which the statement is not true. This is called a counter-example. In order to show by counter-example that two trigonometric expressions are not equivalent to each other, you can substitute a value into each expression. If the two expressions are not equal, then you have proved that they are not equivalent. You can substitute any value, but it is easier to use a value that gives a simple known result. For example, when substituting for θ into any of the trigonometric functions, $\sin\theta$, $\cos\theta$ or $\tan\theta$, choose a value such as 0, π, $\frac{\pi}{2}$, $\frac{\pi}{4}$, etc.

It should be noted that counter-example questions do not always involve trigonometric expressions.

Examples

1   Show by counter-example, that the statement

$\operatorname{cosec}^2\theta \equiv 1 + \sec^2\theta$

is false.

Answer

1   Let $\theta = \dfrac{\pi}{3}$

$\sin\dfrac{\pi}{3} = \dfrac{\sqrt{3}}{2}$ so $\sin^2\dfrac{\pi}{3} = \left(\dfrac{\sqrt{3}}{2}\right)^2 = \dfrac{3}{4}$

LHS $= \operatorname{cosec}^2\theta = \dfrac{1}{\sin^2\theta} = \dfrac{1}{\sin^2\frac{\pi}{3}} = \dfrac{1}{\frac{3}{4}} = \dfrac{4}{3}$

RHS $= 1 + \sec^2\theta = 1 + \dfrac{1}{\cos^2\theta} = 1 + \dfrac{1}{\cos^2\frac{\pi}{3}} = 1 + \dfrac{1}{\frac{1}{4}} = 5$

$\dfrac{4}{3} \neq 5$ so the statement $\operatorname{cosec}^2\theta \equiv 1 + \sec^2\theta$ is false.

2 Show that the statement $|a + b| \equiv |a| + |b|$ is false.

. . . .

Answer

2 Let $a = 1$, $b = -1$

$$\text{LHS} = |1 - 1| = |0| = 0$$

$$\text{RHS} = |1| + |-1| = 1 + 1 = 2$$

$0 \neq 2$ so the statement $|a + b| \equiv |a| + |b|$ is false.

---

3 Show, by counter-example, that the statement
$\sin 2\theta \equiv 2 \sin^2 \theta - \cos \theta$

is false.

> You can let θ be any value but it makes sense to use a value where the sine and cosine of θ are known.

. . . .

Answer

3 Let $\theta = \dfrac{\pi}{2}$

$$\text{LHS} = \sin 2\theta = \sin\left(2 \times \frac{\pi}{2}\right) = \sin \pi = 0$$

> Note that
> $$\sin \frac{\pi}{2} = 1 \text{ and } \cos \frac{\pi}{2} = 1$$

$$\text{RHS} = 2 \sin^2 \theta - \cos \theta = 2 \sin^2 \frac{\pi}{2} - \cos \frac{\pi}{2} = 2 - 0 = 2$$

$0 \neq 2$ so the statement $\sin 2\theta \equiv 2 \sin^2 \theta - \cos \theta$ is false.

---

4 (a) Show, by counter-example, that the statement
$$\sec^2 \theta \equiv 1 - \text{cosec}^2 \theta$$
is false.

  (b) Find all the values of θ in the range $0° \leq \theta \leq 360°$ satisfying
$$3 \,\text{cosec}^2 \theta = 11 - 2 \cot \theta$$

. . . .

Answer

4 (a) Let $\theta = \dfrac{\pi}{4}$

$$\text{LHS} = \sec^2 \theta = \frac{1}{\cos^2 \theta} = \frac{1}{\cos^2 \frac{\pi}{4}} = \frac{1}{\left(\frac{1}{\sqrt{2}}\right)^2} = 2$$

$$\text{RHS} = 1 - \text{cosec}^2 \theta = 1 - \frac{1}{\sin^2 \theta} = 1 - \frac{1}{\sin^2 \frac{\pi}{4}} = 1 - \frac{1}{\left(\frac{1}{\sqrt{2}}\right)^2} = 1 - 2 = -1$$

$2 \neq -1$ so the statement $\sec^2 \theta \equiv 1 - \text{cosec}^2 \theta$ is false.

  (b)
$$3 \,\text{cosec}^2 \theta = 11 - 2 \cot \theta$$
$$3 \,(1 + \cot^2 \theta) = 11 - 2 \cot \theta$$
$$3 + 3 \cot^2 \theta = 11 - 2 \cot \theta$$
$$3 \cot^2 \theta + 2 \cot \theta - 8 = 0$$
$$(3 \cot \theta - 4)(\cot \theta + 2) = 0$$
Hence $\cot \theta = \dfrac{4}{3}$ or $\cot \theta = -2$

> Use $\text{cosec}^2 \theta = 1 + \cot^2 \theta$ to write the equation in terms of $\cot \theta$.

> Remember that the tan function has a period of $180°$ (i.e. π). So from the first angle (e.g. $36.9°$) the other angle can be found by adding $180°$ to it (i.e. $180 + 36.9 = 216.9°$).

tan θ is negative in the second and fourth quadrants, so $\theta = 180 - 26.6 = 153.4°$ or $\theta = 360 - 26.6 = 333.4°$.

Check that all the angles you have found lie in the range $0° \leq \theta \leq 360°$ specified in the question.

$\cot\theta = \dfrac{1}{\tan\theta}$ so $\tan\theta = \dfrac{3}{4}$ giving $\theta = 36.9°$ or $216.9°$

$\cot\theta = \dfrac{1}{\tan\theta}$ so $\tan\theta = -\dfrac{1}{2}$ giving $\theta = 153.4°$ or $333.4°$

---

5 (a) Solve the equation

$$\csc^2 x + \cot^2 x = 5$$

for $0° \leq x \leq 360°$.

(b) (i) Express $4\sin\theta + 3\cos\theta$ in the form $R\sin(\theta + \alpha)$ where $R > 0$ and $0° \leq \alpha \leq 90°$

(ii) Solve the equation

$$4\sin\theta + 3\cos\theta = 2$$

for $0° \leq \theta \leq 360°$, giving your answer correct to the nearest degree.

· ·

Answer

This formula must be memorised as it is not included in the formula booklet.

5 (a) Now $\csc^2 x = 1 + \cot^2 x$

Substituting this into $\csc^2 x + \cot^2 x = 5$

Gives $1 + 2\cot^2 x = 5$

$$\cot^2 x = 2$$

Now $\cot x = \dfrac{1}{\tan x}$ so $\tan x = \dfrac{1}{\cot x}$

Hence, $\tan x = \pm\dfrac{1}{\sqrt{2}}$

$x = 35.3°, 144.7°, 215.3°, 324.7°$

(b) (i) $4\sin\theta + 3\cos\theta \equiv R(\sin\theta\cos\alpha + \cos\theta\sin\alpha)$

$R\cos\alpha = 4$

$R\sin\alpha = 3$

$R = \sqrt{3^2 + 4^2} = 5$

$\dfrac{R\sin\alpha}{R\cos\alpha} = \tan\alpha = \dfrac{3}{4}$

giving $\tan\alpha = \dfrac{3}{4}$ giving $\alpha = 36.87°$

So $4\sin\theta + 3\cos\theta \equiv 5\sin(\theta + 36.87°)$

(ii) $5\sin(\theta + 36.87°) = 2$

$\sin(\theta + 36.87°) = 0.4$

$\theta + 36.87° = 23.58°, 156.42°, 383.58°$

$\theta = 119.55°, 346.71°$

$\theta = 120°, 347°$ (nearest degree)

Active Learning

There are lots of formulae in this topic with some included in the formula book and some which aren't. Use a copy of the formula book to write a list of the formulae you must be familiar with but don't need to remember and also write a list of formulae you may need which are not included in the formula booklet.

Take a picture of the list on your phone and use it to help you to remember those formulae not given.

Test yourself

1. (a) Express $3 \cos \theta + 2 \sin \theta$ in the form $R \cos(\theta - \alpha)$, where $R > 0$ and $0° < \alpha < 90°$. [2]

   (b) Find all the values of θ in the range $0° < \theta < 360°$ satisfying
   $$3 \cos \theta + 2 \sin \theta = 1$$ [3]

2. Showing all your working, find the values of θ in the range $0° \leq \theta \leq 360°$ satisfying the equation $3 \cos 2\theta = 1 - \sin \theta$. [6]

3. Showing all your working, find the values of θ between $0°$ and $360°$ satisfying $4 \sin \theta + 5 \cos \theta = 2$. [5]

4. Show, by counter-example, that the statement
   $$\cos 4\theta \equiv 4 \cos^3 \theta - 3 \cos \theta$$
   is false. [3]

5. Find all values of θ in the range $0° \leq \theta \leq 360°$ satisfying $2 \sec^2 \theta + \tan \theta = 8$. [5]

6. (a) Show, by counter-example, that the statement
   $$\tan 2\theta \equiv \frac{2 \tan \theta}{1 + \tan^2 \theta} \qquad \text{is false.}$$ [3]

   (b) Find all values of θ in the range $0° \leq \theta \leq 360°$ satisfying
   $$2 \sec \theta + \tan^2 \theta = 7.$$ [5]

7. Gwyn wants to turn part of his garden into a circular flower bed. In order to do this, he digs out a shallow circular hole of radius r m and then divides it into two segments by means of a thin plank AB, as shown in the diagram. He plants red roses in the minor segment and white roses in the major segment.

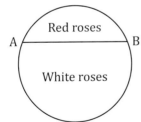

Let the centre of the flower bed be denoted by O. Show that when angle AOB equals 2.6 radians, the area of the flower bed containing white roses is approximately twice the area containing red roses. [5]

Summary

Check you know the following facts:

Radian measure, arc length, area of sector and area of segment

π radians = 180° 2π radians = 360°

$\dfrac{\pi}{2}$ radians = 90° $\dfrac{\pi}{4}$ radians = 45°

$\dfrac{\pi}{3}$ radians = 60° $\dfrac{\pi}{6}$ radians = 30°

The length of an arc making an angle of θ radians at the centre $l = r\theta$

Area of sector making an angle of θ radians at the centre $= \dfrac{1}{2}r^2\theta$

Area of segment $= \dfrac{1}{2}r^2(\theta - \sin\theta)$

sec, cosec and cot

$$\sec\theta = \frac{1}{\cos\theta}$$

$$\operatorname{cosec}\theta = \frac{1}{\sin\theta}$$

$$\cot\theta = \frac{1}{\tan\theta}$$

Trignometric identities

$$\sec^2\theta = 1 + \tan^2\theta$$

$$\operatorname{cosec}^2\theta = 1 + \cot^2\theta$$

$$\sin(A \pm B) = \sin A\cos B \pm \cos A\sin B$$

$$\cos(A \pm B) = \cos A\cos B \mp \sin A\sin B$$

$$\tan(A \pm B) = \frac{\tan A \pm \tan B}{1 \mp \tan A\tan B}$$

Double angle formulae

$$\sin 2A = 2\sin A\cos A$$

$$\cos 2A = \cos^2 A - \sin^2 A$$

$$= 1 - 2\sin^2 A$$

$$= 2\cos^2 A - 1$$

$$\tan 2A = \frac{2\tan A}{1 - \tan^2 A}$$

Important rearrangements of the double angle formulae

$$\sin^2 A = \frac{1}{2}\left(1 - \cos 2A\right)$$

$$\cos^2 A = \frac{1}{2}\left(1 + \cos 2A\right)$$

5 Differentiation

Introduction

This topic builds on the material covered in Topic 7 of the AS book and introduces you to the differentiation of trigonometric functions and more advanced techniques for differentiation.

This topic covers the following:

5.1 Differentiation from first principles for $\sin x$ and $\cos x$

5.2 Using the second derivative to find stationary points and points of inflection

5.3 Differentiation of e^{kx}, a^{kx}, $\sin kx$, $\cos kx$ and $\tan kx$ and related sums, differences and multiples

5.4 The derivative of $\ln x$

5.5 Differentiation using the Chain, Product and Quotient rules

5.6 Differentiation of inverse functions $\sin^{-1} x$, $\cos^{-1} x$, $\tan^{-1} x$

5.7 Connected rates of change and inverse functions

5.8 Differentiation of simple functions defined implicitly

5.9 Differentiation of simple functions and relations defined parametrically

5.10 Constructing simple differential equations

5.1 Differentiation from first principles for sin x and cos x

Differentiation of sin x

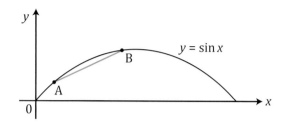

The above graph shows part of the graph $y = \sin x$. Point A has coordinates $(x, \sin x)$ and point B a small distance away has coordinates $(x + \delta x, y + \delta y)$ or $(x + \delta x, \sin(x + \delta x))$.

The formula for
$$\sin(x + \delta x) - \sin x$$
can be found from the formula booklet.

$$\text{Gradient of the line AB is} \quad \frac{\delta y}{\delta x} = \frac{\sin(x + \delta x) - \sin x}{\delta x}$$

A small change in the x-coordinate δx will be accompanied by a small change in the y-coordinate, δy.

The equation of the curve can be used to obtain δy in terms of δx.

$$= \frac{2 \cos\left(\frac{2x + \delta x}{2}\right)\sin\left(\frac{\delta x}{2}\right)}{\delta x}$$

$$= \cos\left(x + \frac{\delta x}{2}\right)\frac{\sin\left(\frac{\delta x}{2}\right)}{\frac{\delta x}{2}}$$

$$\text{Let } \delta x \to 0 \text{ so } \frac{\delta y}{\delta x} \to \frac{dy}{dx} \text{ and } \cos\left(x + \frac{\delta x}{2}\right) \to \cos x \text{ and } \frac{\sin\left(\frac{\delta x}{2}\right)}{\frac{\delta x}{2}} \to 1$$

$$\text{Hence, } \frac{dy}{dx} = \cos x$$

Differentiation of cos x

Using a similar method we can differentiate $\cos x$. Suppose point A has coordinates $(x, \cos x)$ and point B a small distance away has coordinates $(x + \delta x, y + \delta y)$ or $(x + \delta x, \cos(x + \delta x))$.

$$\text{Gradient of the line AB is} \quad \frac{\delta y}{\delta x} = \frac{\cos(x + \delta x) - \cos x}{\delta x}$$

A small change in the x-coordinate δx will be accompanied by a small change in the y-coordinate, δy.

The equation of the curve can be used to obtain δy in terms of δx.

$$= \frac{-2 \sin\left(\frac{2x + \delta x}{2}\right)\sin\left(\frac{\delta x}{2}\right)}{\delta x}$$

$$= -\sin\left(x + \frac{\delta x}{2}\right)\frac{\sin\left(\frac{\delta x}{2}\right)}{\frac{\delta x}{2}}$$

$$\text{Let } \delta x \to 0 \text{ so } \frac{\delta y}{\delta x} \to \frac{dy}{dx} \text{ and } -\sin\left(x + \frac{\delta x}{2}\right) \to -\sin x \text{ and } \frac{\sin\left(\frac{\delta x}{2}\right)}{\frac{\delta x}{2}} \to 1$$

$$\text{Hence, } \frac{dy}{dx} = -\sin x$$

5.2 Using the second derivative to find stationary points and points of inflection

You came across stationary points and points of inflection in Topic 7 of the AS book. Have a look back at pages 136 to 138 of the AS book at the sections on stationary points, points of inflection and the second order derivative.

The second derivative

In order to find the second derivative $\left(\text{i.e. } \frac{d^2y}{dx^2} \text{ or } f''(x)\right)$ you differentiate the first derivative $\left(\text{i.e. } \frac{dy}{dx} \text{ or } f'(x)\right)$.

The second derivative gives the following information about the stationary points:

If $\frac{d^2y}{dx^2}$ or $f''(x) < 0$ the point is a maximum point.

If $\frac{d^2y}{dx^2}$ or $f''(x) > 0$ the point is a minimum point.

If $\frac{d^2y}{dx^2}$ or $f''(x) = 0$ this gives no further information about the nature of the point and further investigation is necessary.

As the first derivative is differentiated again to give the second derivative, the second derivative is the rate of change of the gradient.

Take a look at the following sections of curves.

The graph below shows the curve $y = x^2$.

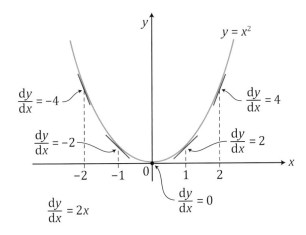

The first derivative of $y = x^2$ is $\frac{dy}{dx} = 2x$. The gradients at the points with x-coordinates $-2, -1, 0, 1$ and 2 are determined and are found to be $-4, -2, 0, 2$ and 4 respectively.

This means that as the x-value increases the gradient is increasing. This means that the second derivative, which represents the change of gradient with x, is increasing, so the value of $\frac{d^2y}{dx^2}$ or $f''(x)$ is positive.

Hence for a minimum point $\frac{d^2y}{dx^2}$ or $f''(x) > 0$ (i.e. the second derivative is positive).

It can be shown in a similar way that for a maximum, the value of $\frac{d^2y}{dx^2}$ or $f''(x)$ is negative.

Active Learning

Write down an equation of a curve that has a single maximum. Find the first derivative and use this to find the x-value of the maximum point. Now find the gradient at various values of x either side of the value of x that corresponds to the maximum.

Look at the way the gradient changes with increasing x-values.

Write down a conclusion for the second derivative.

Concave and convex sections of curves

It is important to be able to distinguish between concave and convex sections of curves. Look at the following, which explains how to tell whether a particular section of a curve is convex or concave.

Concave section of curve.

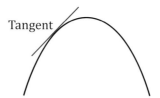

Wherever the tangent is drawn, the tangent line is **above** the curve
Concave = ∩ (think of a frown)

Convex section of curve.

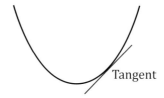

Wherever the tangent is drawn, the tangent line is **below** the curve
Convex = ∪ (think of a smile)

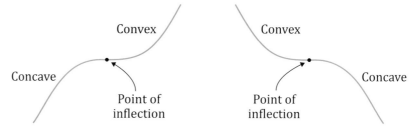

A point of inflection exists when a curve changes from convex to concave or *vice versa*.

In the graph below, as the value of x increases, the graph changes from convex (a smile shape) to concave (a frown shape).
When this happens there is a point of inflection between the two sections and the sign of the second derivative $\left(\frac{d^2y}{dx^2} \text{ or } f''(x)\right)$ changes either side of the point of inflection.

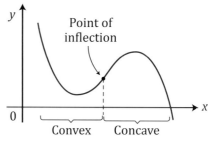

When you think you have found a point of inflection you must always check if $\frac{d^2y}{dx^2}$ changes sign either side of the suspected point of inflection.

For example, here we have a graph of $y = x^3$. You can clearly see there is a point of inflection at the origin. However, you may not have a sketch of the graph as you might be finding the point of inflection so the graph can be drawn.

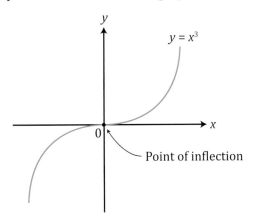

To find the coordinates of the point of inflection you need to find the second derivative and then equate it to zero and then solve the resulting equation.

The first derivative of $y = x^3$ is $\frac{dy}{dx} = 3x^2$ and the second derivative is $\frac{d^2y}{dx^2} = 6x$

At a point of inflection, $\frac{d^2y}{dx^2} = 0$, hence $6x = 0$, so the x-coordinate is $x = 0$. Substituting this value into the equation gives $y = 0$. Hence the point is $(0, 0)$.

Checking the sign of $\frac{d^2y}{dx^2}$ either side of this point we can substitute $x = -0.1$ and $x = 0.1$ in turn into $\frac{d^2y}{dx^2} = 6x$. When $x = -0.1$, $\frac{d^2y}{dx^2}$ is negative and when $x = 0.1$, $\frac{d^2y}{dx^2}$ is positive.

As there is a sign change the point $(0, 0)$ is a point of inflection.

Here is the graph of $y = x^4$.

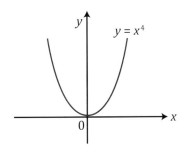

The first derivative of $y = x^4$ is $\frac{dy}{dx} = 4x^3$ and the second derivative is $\frac{d^2y}{dx^2} = 12x^2$.

The second derivative is zero at $(0,0)$.

However, when $x = -0.1$, $\frac{d^2y}{dx^2}$ is positive and when $x = 0.1$, $\frac{d^2y}{dx^2}$ is positive, so this point is not a point of inflection even though $\frac{d^2y}{dx^2} = 0$. This shows the need to investigate the sign of the second derivative before deciding if a point with $\frac{d^2y}{dx^2} = 0$ is a point of inflection.

BOOST

Grade ⇧⇧⇧⇧

Do not assume that if the second derivative is zero the point is a point of inflection. You need to check that the sign of the second derivative changes when x-values which are very near to the point of inflection are substituted into the second derivative.

Finding points of inflection

At a point of inflection, the gradient (i.e. $\frac{dy}{dx}$) does not change sign either side of the point of inflection.

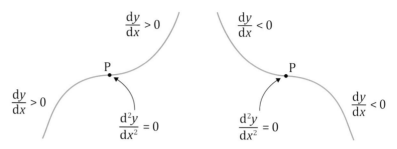

The gradient does not change direction
either side of a point of inflection, P.

At a point of inflection $P(x, y)$, the second derivative $\frac{d^2y}{dx^2}$ or $f''(x) = 0$ and it also changes sign either side of x.

Step by STEP

Find the coordinates of all the points of inflection on the curve

$$y = x^4 - 24x^2 + 11$$

Steps to take

1   Find the first derivative by differentiating the equation of the curve.

2   Find the second derivative by differentiating the first derivative.

3   Set the second derivative to zero and solve the resulting equation for values of x.

4   Check if the sign of the second derivative changes by substituting values for x either side of each suspected point of inflection.

5   Substitute the x-coordinates of the points of inflection into the equation of the curve to find the corresponding y-coordinates.

* *

Answer

$$y = x^4 - 24x^2 + 11$$

$$\frac{dy}{dx} = 4x^3 - 48x$$

$$\frac{d^2y}{dx^2} = 12x^2 - 48$$

At the points of inflection, $\frac{d^2y}{dx^2} = 0$, so $12x^2 - 48 = 0$

Hence, $x^2 - 4 = 0$, so $x^2 = 4$ so $x = \pm 2$

Check that $x = 2$ is a point of inflection by checking the sign of $\frac{d^2y}{dx^2}$ either side of $x = 2$.

When $x = 1$, $\frac{d^2y}{dx^2} = 12 - 48 = -36$ (i.e. negative)

When $x = 3$, $\dfrac{d^2y}{dx^2} = 108 - 48 = 60$ (i.e. positive)

There is a sign change for $\dfrac{d^2y}{dx^2}$, so $x = 2$ is a point of inflection.

Check that $x = -2$ is a point of inflection by checking the sign of $\dfrac{d^2y}{dx^2}$ either side of -2

When $x = -1$, $\dfrac{d^2y}{dx^2} = 12 - 48 = -36$ (i.e. negative)

When $x = -3$, $\dfrac{d^2y}{dx^2} = 108 - 48 = 60$ (i.e. positive)

There is a sign change for $\dfrac{d^2y}{dx^2}$ so $x = -2$ is a point of inflection.

When $x = 2$, $y = 2^4 - 24(2)^2 + 11 = -69$

When $x = -2$, $y = (-2)^4 - 24(-2)^2 + 11 = -69$

So, the points of inflection are at $(-2, -69)$ and $(2, -69)$

Example

1 If $f(x) = x^3 - 3x^2 - 9x + 22$.

   (a) Show that $(x - 2)$ is a factor of $x^3 - 3x^2 - 9x + 22$ and find the points where the curve cuts the x-axis.

   (b) Find the coordinates of the stationary points and the point of inflection on the graph of $f(x)$.

   (c) State the gradient at the point of inflection.

. . . .

Answer

1 (a) $f(2) = 2^3 - 3(2)^2 - 9(2) + 22 = 0$
       (as $f(2) = 0$, $(x - 2)$ is a factor)

   $x^3 - 3x^2 - 9x + 22 = (x - 2)(x^2 + bx + c)$.

   Equating coefficients of x^2 we have $-3 = b - 2$, so $b = -1$

   Equating coefficients of x we have $-9 = c - 2b$, so $c = -11$

   Hence, $x^3 - 3x^2 - 9x + 22 = (x - 2)(x^2 - x - 11)$

   Now at the points where the curve cuts the x-axis, $f(x) = 0$,
   so $(x - 2)(x^2 - x - 11) = 0$

   We can use the formula (or completing the square) to factorise the quadratic part of the cubic. The solutions of $(x - 2)(x^2 - x - 11) = 0$ are

   $$x = 2 \ \text{ or } \ \dfrac{1 \pm \sqrt{45}}{2}$$

   So the curve cuts the x-axis at $x = \dfrac{1 - \sqrt{45}}{2}$, 2 and $\dfrac{1 + \sqrt{45}}{2}$

   (b) $f'(x) = 3x^2 - 6x - 9 = 3(x^2 - 2x - 3) = 3(x + 1)(x - 3)$

   At the stationary points, $f'(x) = 0$ so $3(x + 1)(x - 3) = 0$

   Hence $x = -1, 3$

   When $x = -1, y = 27$ and when $x = 3, y = -5$.

   Hence the stationary points are at $(-1, 27)$ and $(3, -5)$

> You could alternatively use long division to divide $(x - 2)$ into $x^3 - 3x^2 - 9x + 22$ to find the quadratic factor.

> The gradient is zero at the stationary points so $f'(x)$ is equated to zero.

$f''(x) = 6x - 6 = 6(x - 1)$

At $x = -1$, $f''(x) = -12$ so this is a maximum point.

At $x = 3$, $f''(x) = 12$ so this is a minimum point.

At the point of inflection $f''(x) = 0$, so $6(x - 1) = 0$, giving $x = 1$.

At $x = 0$, $f''(x) = -6$

At $x = 2$, $f''(x) = 6$

There is a sign change for $f''(x)$, so $x = 1$ is a point of inflection.

When $x = 1$, $y = 1^3 - 3(1)^2 - 9(1) + 22 = 11$

Hence, the point of inflection is at $(1, 11)$

(c)   At $x = 1$, $f'(1) = 3(1)^2 - 6(1) - 9 = -12$

5.3 Differentiation of e^{kx}, a^{kx}, $\sin kx$, $\cos kx$ and $\tan kx$ and related sums, differences and multiples

The following derivatives will be used in this topic and these will need to be remembered as they are **not** included in the formula booklet:

$$\frac{d(e^{kx})}{dx} = ke^{kx}$$

$$\frac{d(a^{kx})}{dx} = ka^{kx} \ln a$$

$$\frac{d(\sin kx)}{dx} = k \cos kx$$

$$\frac{d(\cos kx)}{dx} = -k \sin kx$$

Note that in all the results shown here, k must be an ordinary number.

The following derivative need not be remembered as it is included in the formula booklet.

$$\frac{d(\tan kx)}{dx} = k \sec^2 kx$$

There are two more derivatives that you need to know and both are included in the formula booklet:

$$\frac{d(\sec x)}{dx} = \sec x \tan x$$

$$\frac{d(\cosec x)}{dx} = -\cosec x \cot x$$

Examples

1   If $y = e^{4x}$, find $\frac{dy}{dx}$.

Answer

1   $y = e^{4x}$, $\qquad\qquad \frac{dy}{dx} = 4e^{4x}$

Use $\frac{d(e^{kx})}{dx} = ke^{kx}$

2 If $y = 3^{2x}$, find $\dfrac{dy}{dx}$.

· ·

Answer

2 $\quad y = 3^{2x}$ $\qquad\qquad \dfrac{dy}{dx} = 2 \times 3^{2x} \ln 3$

Use $\dfrac{d\left(a^{kx}\right)}{dx} = ka^{kx} \ln a$

3 If $y = \sin 2x$, find $\dfrac{dy}{dx}$.

· ·

Answer

3 $\quad y = \sin 2x,$ $\qquad\qquad \dfrac{dy}{dx} = 2 \cos 2x$

Use $\dfrac{d(\sin kx)}{dx} = k \cos kx$

4 If $y = \tan \dfrac{x}{2}$, find $\dfrac{dy}{dx}$.

· ·

Answer

4 $\quad y = \tan \dfrac{x}{2},$ $\qquad\qquad \dfrac{dy}{dx} = \dfrac{1}{2} \sec^2 \dfrac{x}{2}.$

Use $\dfrac{d(\tan kx)}{dx} = k \sec^2 kx$

5.4 The derivative of ln x

The derivative of ln x is not included in the formula booklet and will need to be remembered.

$$\frac{d(\ln x)}{dx} = \frac{1}{x}$$

Examples

1 Differentiate $y = \ln 2x$.

· ·

Answer

1 $\quad y = \ln 2x$

$$y = \ln x + \ln 2$$
$$\frac{dy}{dx} = \frac{1}{x}$$

Remember that when you differentiate a constant term the terms changes to zero.

Hence $\dfrac{d(\ln 2)}{dx} = 0$.

2 Differentiate $y = \ln x^2$.

· ·

Answer

2 $\quad y = \ln x^2$

$$y = \ln x + \ln x$$
$$\frac{dy}{dx} = \frac{1}{x} + \frac{1}{x}$$
$$= \frac{2}{x}$$

5.5 Differentiation using the Chain, Product and Quotient rules

The Chain, Product and Quotient rules are rules that are used to differentiate certain types of function.

The Chain rule

Using the Chain rule, you can differentiate a composite function (sometimes called a function of a function).

Suppose y is a function of u, and u is a function of x, then the Chain rule states:

$$\frac{dy}{dx} = \frac{dy}{du} \times \frac{du}{dx}$$

For example if
$$y = \sin(x^9 + 2 + 1)$$
then $y = \sin u$
where $u = x^9 + 2 + 1$.

Example

1   Differentiate each of the following:

   (a) $y = (4x^2 - 1)^3$

   (b) $y = \sqrt{3x^2 + 1}$

   (c) $y = \ln(5x^2 + 3)$

   (d) $y = \cos(x^3 + x + 5)$

   (e) $y = \tan 3x$

. . . .

Answer

Here the variable u is put equal to the contents inside the bracket.

1   (a) $y = (4x^2 - 1)^3$

Let $u = 4x^2 - 1$ $\left(\dfrac{du}{dx} = 8x\right)$

so that $y = u^3$ $\left(\dfrac{dy}{du} = 3u^2\right)$

Substitute $u = 4x^2 - 1$ back into the equation so that the result contains x only.

Then $\dfrac{dy}{dx} = \dfrac{dy}{du} \times \dfrac{du}{dx} = 3u^2 \times 8x$

The Chain rule is used here. You have to remember it, as it is not given in the formula booklet.

$$= 3(4x^2 - 1)^2 \times 8x$$

$$= 24x(4x^2 - 1)^2$$

   (b) $y = \sqrt{3x^2 + 1} = (3x^2 + 1)^{\frac{1}{2}}$

Let $u = 3x^2 + 1$ $\left(\dfrac{du}{dx} = 6x\right)$

so that $y = u^{\frac{1}{2}}$ $\left(\dfrac{dy}{du} = \dfrac{1}{2}u^{-\frac{1}{2}}\right)$

Remember that
$$u^{-\frac{1}{2}} = \frac{1}{\sqrt{u}}$$ and then substitute u back as $3x^2 + 1$.

Then $\dfrac{dy}{dx} = \dfrac{dy}{du} \times \dfrac{du}{dx} = \dfrac{1}{2}u^{-\frac{1}{2}} \times 6x$

$$= \frac{1}{2\sqrt{3x^2 + 1}} \times 6x$$

$$= \frac{3x}{\sqrt{3x^2 + 1}}$$

(c) $y = \ln (5x^2 + 3)$

Let $u = 5x^2 + 3$ $\left(\dfrac{du}{dx} = 10x\right)$

so that $y = \ln u$ $\left(\dfrac{dy}{du} = \dfrac{1}{u}, \quad \begin{array}{l}\text{see differentiation of the}\\ \ln \text{ function on page 111}\end{array}\right)$

Then $\dfrac{dy}{dx} = \dfrac{dy}{du} \times \dfrac{du}{dx}$

$= \dfrac{1}{u} \times 10x$

$= \dfrac{10x}{5x^2 + 3}$

> Substitute for u in the final result.

(d) $y = \cos (x^3 + x + 5)$

Let $u = x^3 + x + 5$ $\left(\dfrac{du}{dx} = 3x^2 + 1\right)$

so that $y = \cos u$ $\left(\dfrac{dy}{du} = -\sin u\right)$

Then $\dfrac{dy}{dx} = \dfrac{dy}{du} \times \dfrac{du}{dx} = -\sin u \times (3x^2 + 1)$

$= -(3x^2 + 1) \sin (x^3 + x + 5)$

> Substitute for u to obtain the final result.

(e) $y = \tan 3x$

Let $u = 3x$ $\left(\dfrac{du}{dx} = 3\right)$

so that $y = \tan u$ $\left(\dfrac{dy}{du} = \sec^2 u\right)$

Then $\dfrac{dy}{dx} = \dfrac{dy}{du} \times \dfrac{du}{dx} = \sec^2 u \times 3$

$= 3 \sec^2 3x$

> Substitute for u to obtain the final result.

We are able to summarise the work in this example by producing a generalised form of the table given on the first page of this topic.

The results can be obtained in each case by writing:

$u = f(x)$ and noting that $\dfrac{du}{dx} = f'(x)$

$\dfrac{d}{dx}\left(\left(f(x)\right)^n\right) = n\left(f(x)\right)^{n-1} \times f'(x)$

$\dfrac{d}{dx}\left(\left(\sin x\right)^n\right) = n \sin^{n-1} x \times \cos x$

$\dfrac{d}{dx}\left(e^{f(x)}\right) = e^{f(x)} f'(x)$

$\dfrac{d}{dx}\left(e^{x^3+1}\right) = e^{x^3+1} \times 3x^2$

$\dfrac{d}{dx}\left(\ln\left(f(x)\right)\right) = \dfrac{1}{f(x)} \times f'(x)$

$\dfrac{d}{dx}\left(\ln (x^5 + x)\right) = \dfrac{1}{x^5 + x} \times (5x^4 + 1)$

$\dfrac{d}{dx}\left(\sin\left(f(x)\right)\right) = \cos f(x) \times f'(x)$

$\dfrac{d}{dx}(\sin 3x) = \cos 3x \times 3$

$$\frac{d}{dx}\Big(\cos\big(f(x)\big)\Big) = -\sin f(x) \times f'(x)$$

$$\frac{d}{dx}\Big(\cos\big(x^9 + x + 1\big)\Big) = -\sin\big(x^9 + x + 1\big) \times \big(9x^8 +$$

$$\frac{d}{dx}\Big(\tan\big(f(x)\big)\Big) = \sec^2 f(x) \times f'(x)$$

$$\frac{d}{dx}(\tan 5x) = \sec^2 5x \times 5$$

Note

Note the following result in relation to the differentiation of particular logarithmic functions.

If $y = \ln ax$ where a is any constant, then

$$\frac{dy}{dx} = \frac{1}{ax} \times a = \frac{1}{x}$$

which is the same result we obtain if we differentiate $\ln x$.

The result occurs because

$$y = \ln(ax) = \ln a + \ln x \qquad \text{(using the law of logarithms)}$$

and $\quad \frac{d}{dx}(\ln a + \ln x) = \frac{d}{dx}(\ln x) = \frac{1}{x}, \quad$ since $\ln a$ disappears when differentiated.

> $f(x) = ax, \qquad f'(x) = a$

Examples

1 If $y = \ln 3x$, find $\frac{dy}{dx}$.

....

Answer

1 $\quad \dfrac{dy}{dx} = \dfrac{1}{3x} \times 3 = \dfrac{1}{x}$

---

2 If $y = \ln\left(\dfrac{x}{2}\right)$, find $\dfrac{dy}{dx}$.

....

Answer

2 $\quad \dfrac{dy}{dx} = \dfrac{2}{\frac{x}{2}} \times \dfrac{1}{2}$

$\qquad = \dfrac{2}{x} \times \dfrac{1}{2} = \dfrac{1}{x}$

The Product rule

The Product rule is used when you differentiate two different functions of x that are multiplied together.

The Product rule states:

> $f(x)\,g(x)$, means a function f of x multiplied by a function g of x.
>
> f' and g' are the derivatives of f and g respectively.

$$\text{If } y = f(x)\,g(x), \qquad \frac{dy}{dx} = f(x)\,g'(x) + g(x)\,f'(x)$$

You will need to remember the Product rule as it is not given in the formula booklet.

You can remember it as the first term multiplied by the derivative of the second and then added to the second multiplied by the derivative of the first.

Example

1 Differentiate the following with respect to x, simplifying your answer wherever possible.

(a) $x^2 e^{2x}$

(b) $3x^2 \sin 2x$

(c) $(x^3 + 4x^2 - 3)\tan 3x$

· ·

Answer

1 (a) Let $y = x^2 e^{2x}$

$$\frac{dy}{dx} = f(x)\,g'(x) + g(x)\,f'(x)$$

$$= x^2(2e^{2x}) + e^{2x}(2x)$$

$$= 2x^2 e^{2x} + 2xe^{2x} = 2xe^{2x}(x + 1)$$

> All these are the product of one function and another so they need to be differentiated using the Product rule.

> Note that the Product rule must be remembered and note the use of the Chain rule.

(b) Let $y = 3x^2 \sin 2x$

$$\frac{dy}{dx} = f(x)\,g'(x) + g(x)\,f'(x)$$

$$= 3x^2(2\cos 2x) + \sin 2x\,(6x)$$

$$= 6x^2 \cos 2x + 6x \sin 2x$$

$$= 6x(x \cos 2x + \sin 2x)$$

> The derivatives of sin and cos cannot be obtained from the formula booklet so they must be remembered.

(c) Let $y = (x^3 + 4x^2 - 3)\tan 3x$

$$\frac{dy}{dx} = f(x)\,g'(x) + g(x)\,f'(x)$$

$$= (x^3 + 4x^2 - 3)3\sec^2 3x + \tan 3x(3x^2 + 8x)$$

$$= 3(x^3 + 4x^2 - 3)\sec^2 3x + (3x^2 + 8x)\tan 3x$$

> The derivative of tan x is included in the formula booklet and can be looked up.

> The Chain rule is used to differentiate tan $3x$.

The Quotient rule

The Quotient rule is used when you differentiate two different functions of x when one is divided by the other.

The Quotient rule states:

$$\text{If } y = \frac{f(x)}{g(x)} \qquad \frac{dy}{dx} = \frac{f'(x)\,g(x) - f(x)\,g'(x)}{(g(x))^2}$$

> You do not need to remember this formula as it can be looked up in the formula booklet.

Example

1 Differentiate the following with respect to x, simplifying your answer wherever possible.

(a) $\dfrac{x^3 + 2x^2}{x^2 - 1}$

(b) $\dfrac{\sin 2x}{\cos x}$

(c) $\dfrac{e^{5x}}{x^2 + 3}$

Answer

1 (a) Let $y = \dfrac{f(x)}{g(x)} = \dfrac{x^3 + 2x^2}{x^2 - 1}$

$$\dfrac{dy}{dx} = \dfrac{f'(x)\,g(x) - f(x)\,g'(x)}{(g(x))^2}$$

$$= \dfrac{(3x^2 + 4x)(x^2 - 1) - (x^3 + 2x^2)(2x)}{(x^2 - 1)^2}$$

$$= \dfrac{3x^4 - 3x^2 + 4x^3 - 4x - 2x^4 - 4x^3}{(x^2 - 1)^2}$$

$$= \dfrac{x^4 - 3x^2 - 4x}{(x^2 - 1)^2} = \dfrac{x(x^3 - 3x - 4)}{(x^2 - 1)^2}$$

(b) Let $y = \dfrac{f(x)}{g(x)} = \dfrac{\sin 2x}{\cos x}$

$$\dfrac{dy}{dx} = \dfrac{f'(x)\,g(x) - f(x)\,g'(x)}{(g(x))^2}$$

> The Chain rule is used to differentiate $\sin 2x$.

$$= \dfrac{2\cos 2x(\cos x) - \sin 2x(-\sin x)}{\cos^2 x}$$

$$= \dfrac{2\cos 2x \cos x + \sin 2x \sin x}{\cos^2 x}$$

(c) Let $y = \dfrac{f(x)}{g(x)} = \dfrac{e^{5x}}{x^2 + 3}$

$$\dfrac{dy}{dx} = \dfrac{f'(x)\,g(x) - f(x)\,g'(x)}{(g(x))^2}$$

> The Chain rule is used to differentiate e^{5x}.

$$= \dfrac{5e^{5x}(x^2 + 3) - e^{5x}(2x)}{(x^2 + 3)^2}$$

$$= \dfrac{5x^2 e^{5x} + 15e^{5x} - 2xe^{5x}}{(x^2 + 3)^2}$$

$$= \dfrac{e^{5x}(5x^2 - 2x + 15)}{(x^2 + 3)^2}$$

5.6 Differentiation of inverse functions $\sin^{-1}x$, $\cos^{-1}x$, $\tan^{-1}x$

Remember that $y = \sin^{-1} x$ is equivalent to $\sin y = x$, that $y = \cos^{-1} x$ is equivalent to $\cos y = x$ and that $y = \tan^{-1} x$ is equivalent to $\tan y = x$.

To differentiate these inverse trigonometric functions, we note that

$$\dfrac{dy}{dx} = \dfrac{1}{\left(\dfrac{dx}{dy}\right)}$$

Proving $\dfrac{d}{dx}(\sin^{-1}x) = \dfrac{1}{\sqrt{1-x^2}}$

Given that $\quad y = \sin^{-1}x,$

$\quad\quad\quad\quad \sin y = x$

Differentiate with respect to y

$$\cos y = \frac{dx}{dy}$$

Then $\quad \dfrac{dy}{dx} = \dfrac{1}{\left(\dfrac{dx}{dy}\right)} = \dfrac{1}{\cos y}$

$$= \frac{1}{\pm\sqrt{1-\sin^2 y}}$$

$$= \frac{1}{\pm\sqrt{1-x^2}}$$

Remember that
$\sin^2 x + \cos^2 x = 1$

$\sin y = x$

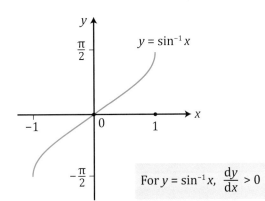

For $y = \sin^{-1}x,\ \dfrac{dy}{dx} > 0$

Since $\dfrac{dy}{dx} > 0,\quad \dfrac{dy}{dx} = \dfrac{1}{\sqrt{1-x^2}}$

Proving $\dfrac{d}{dx}(\cos^{-1}x) = \dfrac{-1}{\sqrt{1-x^2}}$

Given that $\quad y = \cos^{-1}x,$

$\quad\quad\quad\quad \cos y = x$

Differentiate with respect to y

$$-\sin y = \frac{dx}{dy}$$

Then $\quad \dfrac{dy}{dx} = \dfrac{1}{\left(\dfrac{dx}{dy}\right)} = \dfrac{-1}{\sin y}$

$$= \frac{-1}{\pm\sqrt{1-\cos^2 y}}$$

Remember that
$\sin^2 x + \cos^2 x = 1$

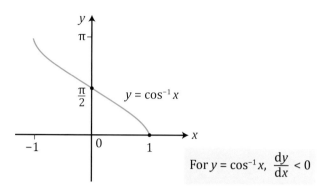

For $y = \cos^{-1}x$, $\dfrac{dy}{dx} < 0$

Since $\dfrac{dy}{dx} < 0$, $\dfrac{dy}{dx} = \dfrac{-1}{\sqrt{1-x^2}}$

We leave the derivation of $\dfrac{d}{dx}(\tan^{-1}x) = \dfrac{1}{1+x^2}$ as a later exercise.

To summarise

$$\frac{d(\sin^{-1}x)}{dx} = \frac{1}{\sqrt{1-x^2}}$$

$$\frac{d(\cos^{-1}x)}{dx} = -\frac{1}{\sqrt{1-x^2}}$$

$$\frac{d(\tan^{-1}x)}{dx} = \frac{1}{1+x^2}$$

These results can be generalised by taking $u = f(x)$ and using the Chain rule.

$$\frac{d}{dx}\left(\sin^{-1}\left(f(x)\right)\right) = \frac{1}{\sqrt{1-(f(x))^2}} \times f'(x)$$

$$\frac{d}{dx}\left(\sin^{-1}(x^3)\right) = \frac{1}{\sqrt{1-(x^3)^2}} \times 3x^2$$

$$= \frac{3x^2}{\sqrt{1-x^6}}$$

$$\frac{d}{dx}\left(\cos^{-1}\left(f(x)\right)\right) = \frac{-1}{\sqrt{1-(f(x))^2}} \times f'(x)$$

$$\frac{d}{dx}\left(\cos^{-1}(5x)\right) = \frac{-1}{\sqrt{1-(5x)^2}} \times 5$$

$$= \frac{-5}{\sqrt{1-25x^2}}$$

$$\frac{d}{dx}\left(\tan^{-1}\left(f(x)\right)\right) = \frac{1}{1+(f(x))^2} \times f'(x)$$

$$\frac{d}{dx}\left(\tan(x^4)\right) = \frac{1}{1+(x^4)^2} \times 4x^3 = \frac{4x^3}{1+x^8}$$

Examples

1   Differentiate each of the following

(a) $y = \cos^{-1}4x$

(b) $y = \sin^{-1}\dfrac{x}{2}$

(c) $y = 3\tan^{-1}2x$

Answers

1 (a) $y = \cos^{-1} 4x$

$$\frac{dy}{dx} = -\frac{4}{\sqrt{1 - (4x)^2}} = -\frac{4}{\sqrt{1 - 16x^2}}$$

(b) $y = \sin^{-1}\frac{x}{2}$

$$\frac{dy}{dx} = \frac{\frac{1}{2}}{\sqrt{1 - \left(\frac{x}{2}\right)^2}} = \frac{1}{2\sqrt{1 - \frac{x^2}{4}}} = \frac{1}{\sqrt{4 - x^2}}$$

(c) $y = 3\tan^{-1} 2x$

$$\frac{dy}{dx} = \frac{3 \times 2}{1 + (2x)^2} = \frac{6}{1 + 4x^2}$$

---

2 (a) Differentiate each of the following with respect to x, simplifying your answer wherever possible:

(i) $\sqrt{2 + 5x^3}$

(ii) $x^2 \sin 3x$

(iii) $\dfrac{e^{2x}}{x^4}$

(b) By first writing $y = \tan^{-1} x$ as $x = \tan y$, find $\dfrac{dy}{dx}$ in terms of x.

Answer

2 (a) (i) $y = \sqrt{2 + 5x^3}$

$$y = (2 + 5x^3)^{\frac{1}{2}}$$

$$\frac{dy}{dx} = \left(\frac{1}{2}\right)(2 + 5x^3)^{-\frac{1}{2}}(15x^2)$$

$$= \frac{15x^2}{2\sqrt{2 + 5x^3}}$$

> You have to recognise that this is a function of a function and therefore needs to be differentiated using the Chain rule.
>
> $y = (f(x))^n$
> with $f(x) = (2 + 5x^3)$, $n = \dfrac{1}{2}$

(ii) $y = x^2 \sin 3x$

$$\frac{dy}{dx} = x^2\, 3\cos 3x + \sin 3x \times 2x$$

$$= 3x^2 \cos 3x + 2x \sin 3x = x(3x \cos 3x + 2\sin 3x)$$

> The Product rule is used here.

(iii) $y = \dfrac{e^{2x}}{x^4}$

$$\frac{dy}{dx} = \frac{f'(x)\,g(x) - f(x)\,g'(x)}{(g(x))^2}$$

$$= \frac{2e^{2x}(x^4) - e^{2x}(4x^3)}{(x^4)^2}$$

> This is the Quotient rule and can be obtained from the formula booklet.

$$= \frac{2x^4 e^{2x} - 4x^3 e^{2x}}{x^8}$$

The top and bottom of this fraction is divided by x^3.

$$= \frac{2x^3 e^{2x}(x - 2)}{x^8}$$

$$= \frac{2e^{2x}(x - 2)}{x^5}$$

$\sec^2 y = 1 + \tan^2 y$ (note: this should be remembered from Topic 1 as it is not given in the formula booklet).

(b) $x = \tan y$

$$\frac{dx}{dy} = \sec^2 y$$

$$= 1 + \tan^2 y$$

$$= 1 + x^2$$

Now $\dfrac{dy}{dx} = \dfrac{1}{\left(\frac{dx}{dy}\right)} = \dfrac{1}{1 + x^2}$

---

3 (a) Differentiate each of the following with respect to x, simplifying your answer wherever possible.

(i) $(7 + 2x)^{13}$

(ii) $\sin^{-1} 5x$

(iii) $x^3 e^{4x}$

(b) By first writing $\tan x = \dfrac{\sin x}{\cos x}$, show that $\dfrac{d}{dx}(\tan x) = \sec^2 x$

· ·

Answer

3 (a) (i) $y = (7 + 2x)^{13}$

$$\frac{dy}{dx} = 13(7 + 2x)^{12}(2)$$

$$= 26(7 + 2x)^{12}$$

From the formula booklet we have derivative of
$$\sin^{-1} x = \frac{1}{\sqrt{1 - x^2}}$$

(ii) $y = \sin^{-1} 5x$

$$d(\sin^{-1} x) = \frac{1}{\sqrt{1 - x^2}}$$

Use the Chain rule with $u = 5x$, or use

$$\frac{d}{dx}\left(\sin^{-1} g(x)\right) = \frac{g'(x)}{\sqrt{1 - (g(x))^2}}$$

$$\frac{d(\sin^{-1} 5x)}{dx} = \frac{5}{\sqrt{1 - (5x)^2}}$$

$$= \frac{5}{\sqrt{1 - 25x^2}}$$

This is a product, so the Product rule is used. The Product rule is not in the formula booklet.

(iii) $y = x^3 e^{4x}$

$$\frac{dy}{dx} = x^3(4e^{4x}) + e^{4x}(3x^2)$$

$$= 4x^3 e^{4x} + 3x^2 e^{4x}$$

$$= x^2 e^{4x}(4x + 3)$$

(b) $\dfrac{d}{dx}(\tan x) = \dfrac{d}{dx}\left(\dfrac{\sin x}{\cos x}\right)$

$= \dfrac{\cos x\,(\cos x) - \sin x\,(-\sin x)}{\cos^2 x}$

$= \dfrac{\cos^2 x + \sin^2 x}{\cos^2 x}$

$= \dfrac{1}{\cos^2 x}$

$= \sec^2 x$

> This is a quotient, so the Quotient rule is used to find the derivative.

> $\sin^2 x + \cos^2 x = 1$
> You need to remember this from your AS studies as it is not included in the formula booklet.

---

4 Differentiate each of the following with respect to x, simplifying your answer wherever possible.

(a) $\tan^{-1} 5x$

(b) $\ln (3x^2 + 5x - 1)$

(c) $e^{3x} \cos x$

(d) $\dfrac{1 + \sin x}{1 - \sin x}$

· ·

Answer

4 (a) $y = \tan^{-1} 5x$

$\dfrac{dy}{dx} = \dfrac{5}{1 + (5x)^2}$

$= \dfrac{5}{1 + 25x^2}$

> Use $\dfrac{d(\tan^{-1} x)}{dx} = \dfrac{1}{1 + x^2}$ obtained from the formula booklet. Then either use the Chain rule or remember that
> $\dfrac{d(\tan^{-1} ax)}{dx} = \dfrac{a}{1 + (ax)^2}$

(b) $y = \ln (3x^2 + 5x - 1)$

$\dfrac{dy}{dx} = \dfrac{6x + 5}{3x^2 + 5x - 1}$

> To differentiate a ln function, differentiate the function and then divide by the original function.

(c) $y = e^{3x} \cos x$

$\dfrac{dy}{dx} = e^{3x}(-\sin x) + \cos x(3e^{3x})$

$= -e^{3x} \sin x + 3e^{3x} \cos x$

$= e^{3x}(-\sin x + 3 \cos x)$

> This represents a product so the Product rule is used.

> The Product rule is not included in the formula booklet.

(d) $y = \dfrac{1 + \sin x}{1 - \sin x}$

$\dfrac{dy}{dx} = \dfrac{(\cos x)(1 - \sin x) - (1 + \sin x)(-\cos x)}{(1 - \sin x)^2}$

$= \dfrac{2\cos x}{(1 - \sin x)^2}$

> This is a quotient so the Quotient rule is used. You can find it in the formula booklet. The formula for the Quotient rule is shown on p.115.

> You must remember the derivatives of $\sin x$ and $\cos x$ as they are not given in the formula booklet.

5   Differentiate each of the following with respect to x, simplifying your answers wherever possible.

(a) $(1 + 3x)^{11}$

(b) $\ln(3 + x^3)$

(c) $\dfrac{\cos x}{1 - \sin x}$

(d) $\tan^{-1}(4x)$

(e) $x^4 \tan x$

· ·

Answer

5   (a)   Let $y = (1 + 3x)^{11}$

$\dfrac{dy}{dx} = 11(1 + 3x)^{10}(3)$

$\quad\ = 33(1 + 3x)^{10}$

Use $\dfrac{d}{dx}\big((f(x))^n\big) = n(f(x))^{n-1} \times f'(x)$

with $f(x) = 1 + 3x, \ n = 11$

(b)   Let $y = \ln(3 + x^3)$

$\dfrac{dy}{dx} = \dfrac{3x^2}{3 + x^3}$

Use $\dfrac{d(\ln(f(x)))}{dx} = \dfrac{f(x)}{f'(x)}$

(c)   Let $y = \dfrac{\cos x}{1 - \sin x}$

This is a quotient, so the Quotient rule is used to differentiate.

$\dfrac{dy}{dx} = \dfrac{(-\sin x)(1 - \sin x) - (\cos x)(-\cos x)}{(1 - \sin x)^2}$

$\quad\ = \dfrac{-\sin x + \sin^2 x + \cos^2 x}{(1 - \sin x)^2}$

Use $\sin^2 x + \cos^2 x = 1$.

$\quad\ = \dfrac{\sin^2 x + \cos^2 x - \sin x}{(1 - \sin x)^2}$

Divide top and bottom by $(1 - \sin x)$.

$\quad\ = \dfrac{1 - \sin x}{(1 - \sin x)^2}$

$\quad\ = \dfrac{1}{1 - \sin x}$

(d)   Let $y = \tan^{-1}(4x)$

Use the formula:
$\dfrac{d(\tan^{-1}x)}{dx} = \dfrac{1}{1 + x^2}$
which can be looked up in the formula booklet
and let $u = 4x$.

$\dfrac{dy}{dx} = \dfrac{4}{1 + (4x)^2}$

$\quad\ = \dfrac{4}{1 + 16x^2}$

(e)   Let $y = x^4 \tan x$

$\dfrac{dy}{dx} = f(x)\,g'(x) + f'(x)\,g(x)$

$\quad\ = x^4 \sec^2 x + (\tan x) \times (4x^3)$

$\quad\ = x^4 \sec^2 x + 4x^3 \tan x$

$\quad\ = x^3(x \sec^2 x + 4 \tan x)$

5.7 Connected rates of change and inverse functions

Connected rates of change

Suppose we are asked to find $\dfrac{dV}{dt}$ but do not have an equation connecting V and t to differentiate directly. If we had the connected equations for A in terms of t and V in terms of A, then we can differentiate each of these to find $\dfrac{dA}{dt}$ and $\dfrac{dV}{dA}$ respectively.

The Chain rule can now be used in the following way to find $\dfrac{dV}{dt}$.

$$\frac{dV}{dt} = \frac{dA}{dt} \times \frac{dV}{dA}$$

This method is easier to understand by looking at the following examples.

Examples

1. When a circular metal disc is heated its radius increases at a rate of 0.005 mm s⁻¹. Find the rate at which the area is increasing when the radius of the disc is 100 mm. Give your answer to 2 significant figures.

Answer

1. We know $\dfrac{dr}{dt} = 0.005$ and we need to find $\dfrac{dA}{dt}$

   Now $$\frac{dA}{dt} = \frac{dr}{dt} \times \frac{dA}{dr}$$

   To determine $\dfrac{dA}{dr}$ we need an equation connecting A and r.

   We can use, $A = \pi r^2$, so $\dfrac{dA}{dr} = 2\pi r$

   $$\frac{dA}{dt} = \frac{dr}{dt} \times \frac{dA}{dr} = 0.005 \times 2\pi r$$

   Now as $r = 100$, $\dfrac{dA}{dt} = 0.005 \times 2\pi \times 100 = 3.1 \text{ mm}^2 \text{ s}^{-1}$ (2 s.f.)

---

2. At time t seconds, the surface area of a cube is A cm² and the volume is V cm³.

   If the surface area of the cube expands at the constant rate of 2 cm² s⁻¹,

   show that $\dfrac{dV}{dt} = \dfrac{1}{2}\sqrt{\dfrac{A}{6}}$

Answer

2. Let x be the length of the side of the cube in cm.

   We can now form the following equation connecting the surface area A with the length of side x.

   $$A = 6x^2$$

   We can also form an equation connecting V and x.

   $$V = x^3$$

Write down what you know $\left(\text{i.e. } \dfrac{dA}{dt}\right)$ and think about what you must multiply it by to give $\dfrac{dV}{dt}$. Here you need a dV on the top and dA on the bottom so that it will cancel with the dA on the top.

We also know that $\dfrac{dA}{dt} = 2$ as this is given in the question.

Now, $$\dfrac{dV}{dt} = \dfrac{dA}{dt} \times \dfrac{dV}{dA}$$

We already know $\dfrac{dA}{dt}$ so we need an equation connecting V and A so that this can be differentiated to find $\dfrac{dV}{dA}$.

We need to eliminate x between the two equations $A = 6x^2$ and $V = x^3$

From the first equation, $x = \sqrt{\dfrac{A}{6}}$ and substituting this into the second equation

we obtain $$V = \left(\sqrt{\dfrac{A}{6}}\right)^3 = \left(\dfrac{A}{6}\right)^{\frac{3}{2}} = \dfrac{1}{6\sqrt{6}}A^{\frac{3}{2}} \quad \left(\text{Note } 6^{\frac{3}{2}} = 6\sqrt{6}\right)$$

Differentiating we obtain $$\dfrac{dV}{dA} = \dfrac{3}{2} \times \dfrac{1}{6\sqrt{6}}A^{\frac{1}{2}} = \dfrac{1}{4}\sqrt{\dfrac{A}{6}}$$

Now, $$\dfrac{dV}{dt} = \dfrac{dA}{dt} \times \dfrac{dV}{dA}$$
$$= 2 \times \dfrac{1}{4}\sqrt{\dfrac{A}{6}} = \dfrac{1}{2}\sqrt{\dfrac{A}{6}}$$

Differentiation using inverse functions

In some problems you might want to find the derivative $\dfrac{dy}{dx}$ but can only find $\dfrac{dx}{dy}$.

It is easy to convert from one to the other using the following result:
$$\dfrac{dy}{dx} = \dfrac{1}{\dfrac{dx}{dy}}$$

Example

1 A spherical balloon is filled with helium gas at the rate of $20\ \text{cm}^3\ \text{s}^{-1}$. Find the rate of increase of the radius when the radius is 10 cm. Give your answer to 2 significant figures.

Answer

1 We know $\dfrac{dV}{dt} = 20$ and are required to find $\dfrac{dr}{dt}$.

$$\dfrac{dr}{dt} = \dfrac{dV}{dt} \times \dfrac{dr}{dV}$$

Now $$\dfrac{dr}{dV} = \dfrac{1}{\dfrac{dV}{dr}}$$

$V = \dfrac{4}{3}\pi r^3$, so $\dfrac{dV}{dr} = 4\pi r^2$

Hence $$\dfrac{dr}{dt} = \dfrac{dV}{dt} \times \dfrac{dr}{dV} = 20 \times \dfrac{1}{4\pi(10)^2} = 0.016\ \text{cm s}^{-1}$$

5.8 Differentiation of simple functions defined implicitly

Finding $\dfrac{dy}{dx}$ in terms of both x and y is called **implicit differentiation**.

Here are some of the basics:

$$\frac{d(3x)^2}{dx} = 6x$$

and

$$\frac{d(6y^3)}{dx} = 18y^2 \times \frac{dy}{dx}$$

$$\frac{d(y)}{dx} = 1 \times \frac{dy}{dx}$$

$$\frac{d(x^2y^3)}{dx} = (x^2)\left(3y^2 \times \frac{dy}{dx}\right) + (y^3)(2x)$$

$$= 3x^2y^2\frac{dy}{dx} + 2xy^3$$

> Terms involving x or constant terms are differentiated as normal.

> Differentiate with respect to y and then multiply the result by $\frac{dy}{dx}$.

> Because there are two terms here, the Product rule is used. Notice the need to include $\frac{dy}{dx}$ when the term involving y is differentiated.

Example

1 Find $\dfrac{dy}{dx}$ for the equation $2x^2 + xy + y^3 = 15$

> The Product rule needs to be used to differentiate the middle term xy.

Answer

1   Differentiating with respect to x we obtain:

$$4x + (x)(1)\frac{dy}{dx} + (y)(1) + 3y^2\frac{dy}{dx} = 0$$

$$4x + x\left(\frac{dy}{dx}\right) + y + 3y^2\frac{dy}{dx} = 0$$

$$x\left(\frac{dy}{dx}\right) + 3y^2\frac{dy}{dx} = -y - 4x$$

$$\frac{dy}{dx}(x + 3y^2) = -y - 4x$$

$$\frac{dy}{dx} = \frac{-y - 4x}{x + 3y^2}$$

> Collect all the terms containing $\frac{dy}{dx}$ on one side of the equation and all the other terms on the other.

> Take $\frac{dy}{dx}$ out of the brackets as a factor.

BOOST
Grade ⇧⇧⇧⇧

It is a common mistake for students to forget to differentiate the right-hand side of the equation, especially if it is just a number.

5.9 Differentiation of simple functions and relations defined parametrically

x and y can be defined in terms of a third variable called a parameter. The equation of a curve can be expressed in parametric form by using $x = f(t), y = g(t)$ where t is the parameter being used.

The formulae for differentiating parametric forms are:

$$\frac{dy}{dx} = \frac{\frac{dy}{dt}}{\frac{dx}{dt}} = \frac{dy}{dt} \times \frac{dt}{dx}$$

and

$$\frac{d^2y}{dx^2} = \frac{\frac{d}{dt}\left(\frac{dy}{dx}\right)}{\frac{dx}{dt}} = \frac{d}{dt}\left(\frac{dy}{dx}\right) \times \frac{dt}{dx}$$

An example of the parametric equation of a curve is

$$x = t^2, \quad y = 2t$$

To determine the gradient of the tangent to a curve defined parametrically we use the formula for $\dfrac{dy}{dx}$ given above.

> Remember to invert $\frac{dx}{dt}$ in this equation.

$$\frac{dx}{dt} = 2t \quad \text{so} \quad \frac{dt}{dx} = \frac{1}{2t}$$

$$\frac{dy}{dt} = 2$$

> If a value of t is known then this can be substituted in so that the gradient can be expressed as a number.

$$\frac{dy}{dx} = \frac{dy}{dt} \times \frac{dt}{dx} = 2 \times \frac{1}{2t} = \frac{1}{t}$$

Examples

1 Given that $x = 2t^2 + 1, y = \dfrac{4t + 1}{t + 1}$, find

(a) $\dfrac{dy}{dt}$

(b) $\dfrac{dy}{dx}$

* *

Answer

1 (a) $\dfrac{dy}{dt} = \dfrac{4(t + 1) - (4t + 1)(1)}{(t + 1)^2}$

> The Quotient rule is used here.

$$= \frac{4t + 4 - 4t - 1}{(t + 1)^2}$$

$$= \frac{3}{(t + 1)^2}$$

(b) $\dfrac{dx}{dt} = 4t$

$$\dfrac{dy}{dx} = \dfrac{dy}{dt} \times \dfrac{dt}{dx}$$

$$= \dfrac{3}{(t+1)^2} \times \dfrac{1}{4t}$$

$$= \dfrac{3}{4t(t+1)^2}$$

$\dfrac{dx}{dt}$ is inverted to give $\dfrac{dt}{dx}$.

---

2 Given that $x = \ln t$, $y = 3t^4 - 2t$,

(a) find an expression for $\dfrac{dy}{dx}$ in terms of t,

(b) find the value of $\dfrac{d^2y}{dx^2}$ when $t = \dfrac{1}{2}$

. . . .

Answer

2 (a) $x = \ln t$

$$\dfrac{dx}{dx} = \dfrac{1}{t}$$

$y = 3t^4 - 2t$

$$\dfrac{dy}{dt} = 12t^3 - 2$$

$$\dfrac{dy}{dx} = \dfrac{dy}{dt} \times \dfrac{dt}{dx}$$

$$= (12t^3 - 2)t$$

$$= 12t^4 - 2t$$

$$= 2t(6t^3 - 1)$$

$\dfrac{dt}{dx} = \dfrac{1}{\frac{1}{t}} = t$

(b) $\dfrac{d^2y}{dx^2} = \dfrac{d}{dt}\left(12t^4 - 2t\right) \times \dfrac{dt}{dx}$

$$= (48t^3 - 2)t$$

$$= 2t(24t^3 - 1)$$

When $t = \dfrac{1}{2}$, $\dfrac{d^2y}{dx^2} = 2\left(\dfrac{1}{2}\right)\left(24 \times \left(\dfrac{1}{2}\right)^3 - 1\right) = 1(3 - 1) = 2$

The first derivative is differentiated again to find the second derivative.

---

3 (a) Given that

$x^4 + 3x^2y - 2y^2 = 15,$

find an expression for $\dfrac{dy}{dx}$ in terms of x and y.

(b) Given that $x = \ln t$, $y = t^3 - 7t$,

(i) find an expression for $\dfrac{dy}{dx}$ in terms of t

(ii) find the value of $\dfrac{d^2y}{dx^2}$ when $t = \dfrac{1}{3}$.

Answer

3   (a) Differentiating implicitly with respect to x, we obtain

$$4x^3 + 3x^2\frac{dy}{dx} + 6xy - 4y\frac{dy}{dx} = 0$$

$$3x^2\frac{dy}{dx} - 4y\frac{dy}{dx} = -4x^3 - 6xy$$

$$\frac{dy}{dx}\left(3x^2 - 4y\right) = -4x^3 - 6xy$$

$$\frac{dy}{dx} = \frac{-4x^3 - 6xy}{3x^2 - 4y}$$

> Multiply the top and bottom by -1.

$$= \frac{4x^3 + 6xy}{4y - 3x^2}$$

(b) (i) $x = \ln t$

$$\frac{dx}{dt} = \frac{1}{t}$$

$$y = t^3 - 7t$$

$$\frac{dy}{dt} = 3t^2 - 7$$

$$\frac{dy}{dx} = \frac{dy}{dt} \times \frac{dt}{dx}$$

$$= (3t^2 - 7)t$$

$$= 3t^3 - 7t$$

(ii) $$\frac{d^2y}{dx^2} = \frac{\frac{d}{dt}\left(\frac{dy}{dx}\right)}{\frac{dx}{dt}}$$

> The Chain rule is used here to find the second derivative.

$$\frac{d^2y}{dx^2} = \frac{d}{dt}\left(\frac{dy}{dx}\right) \times \frac{dt}{dx}$$

$$\frac{d^2y}{dx^2} = \frac{d}{dt}\left(3t^3 - 7t\right) \times t$$

$$= (9t^2 - 7)t$$

> As $\frac{dx}{dt} = \frac{1}{t}$, $\frac{dt}{dx} = \frac{1}{\frac{dx}{dt}} = \frac{1}{\frac{1}{t}} = t$

$$= 9t^3 - 7t$$

When $x = \frac{1}{3}$, $\frac{d^2y}{dx^2} = 9\left(\frac{1}{3}\right)^3 - 7\left(\frac{1}{3}\right) = -2$

---

4   The curve $y = ax^4 + bx^3 + 18x^2$ has a point of inflection at $(1, 11)$.

(a) Show that $2a + b + 6 = 0$.

(b) Find the values of the constants a and b and show that the curve has another point of inflection at $(3, 27)$.

(c) Sketch the curve, identifying all the stationary points including their nature.

Answer

4 (a) When $x = 1$, $y = 11$, so $11 = a(1)^4 + b(1)^3 + 18(1)^2$

$$11 = a + b + 18$$

$$0 = a + b + 7 \qquad (1)$$

This is not the required answer so there must be another equation we can find.

We can use the point of inflection to produce another equation.

$$\frac{dy}{dx} = 4ax^3 + 3bx^2 + 36x$$

$$\frac{d^2y}{dx^2} = 12ax^2 + 6bx + 36$$

At a point of inflection, $\frac{d^2y}{dx^2} = 0$, hence $12ax^2 + 6bx + 36 = 0$

We know the x-coordinate of the point of inflection, so substituting this in we obtain:

$$12a(1)^2 + 6b(1) + 36 = 0$$

Dividing this equation by 6, we obtain $2a + b + 6 = 0 \qquad (2)$

(b) Solving equations (1) and (2) simultaneously we have

$$(2) - (1) \text{ gives } a = 1$$

Substituting $a = 1$ into either equation gives $b = -8$

Now $\frac{d^2y}{dx^2} = 12ax^2 + 6bx + 36$ so substituting these values in gives

$$\frac{d^2y}{dx^2} = 12x^2 - 48x + 36$$

At points of inflection, $\frac{d^2y}{dx^2} = 0$,   so $12x^2 - 48x + 36 = 0$

Dividing through by 12, we obtain $x^2 - 4x + 3 = 0$

Factorising, we obtain $(x - 3)(x - 1) = 0$

Hence $x = 1$ or 3

We are told there is a point of inflection at $x = 1$ so finding the y-coordinate for $x = 3$, we have $y = x^4 - 8x^3 + 18x^2$
so $y = 3^4 - 8(3)^3 + 18(3)^2 = 27$

Hence there is another point of inflection at $(3, 27)$.

When $x = 2$, $\frac{d^2y}{dx^2} = 4 - 8 + 3 = -1$ (i.e. negative)

When $x = 4$, $\frac{d^2y}{dx^2} = 16 - 16 + 3 = 3$ (i.e. positive)

There is a sign change for $\frac{d^2y}{dx^2}$ so $x = 3$ is a point of inflection.

(c) $y = x^4 - 8x^3 + 18x^2$

$$\frac{dy}{dx} = 4x^3 - 24x^2 + 36x$$

We have found the points of inflection and we now need to find any stationary points.

At the stationary points, $\frac{dy}{dx} = 0$

$$4x^3 - 24x^2 + 36x = 0$$

$$x^3 - 6x^2 + 9x = 0$$

$$x(x^2 - 6x + 9) = 0$$

$$x(x - 3)(x - 3) = 0$$

Hence $x = 0$ or 3

When $x = 0$, $y = 0$ and $\dfrac{d^2y}{dx^2} = 36$, and since this is positive it is a minimum point.

These points can now be used to sketch the following graph:

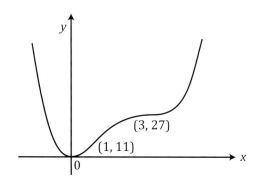

---

5   **(a)**   Differentiate $\cos x$ from first principles.

(b)   Differentiate the following with respect to x, simplifying your answer as far as possible:

(i) $\dfrac{3x^2}{x^3 + 1}$

(ii) $x^3 \tan 3x$

. . . .

Answer

5   **(a)**   See page 104 of this book.

(b)   **(i)**   Using the Quotient rule

The Quotient rule is included in the formula booklet.

$$\text{Let}\quad y = \frac{f(x)}{g(x)} = \frac{3x^2}{x^3 + 1}$$

$$\frac{dy}{dx} = \frac{f'(x)\,g(x) - f(x)\,g'(x)}{(g(x))^2}$$

$$= \frac{6x(x^3 + 1) - (3x^2)(3x^2)}{(x^3 + 1)^2}$$

$$= \frac{6x^4 + 6x - 9x^4}{(x^3 + 1)^2}$$

$$= \frac{6x - 3x^4}{(x^3 + 1)^2} = \frac{3x(2 - x^3)}{(x^3 + 1)^2}$$

(ii)   Let $y = x^3 \tan 3x$

The Product rule is not included in the formula booklet so will need to be remembered.

Using the Product rule:

$$\frac{dy}{dx} = 3x^2 \tan 3x + 3x^3 \sec^2 3x$$

6   Differentiate each of the following with respect to x, simplifying your answer wherever possible.

(a) $\ln(\cos x)$

(b) $e^{6x}(3x - 2)^4$

(c) Show that $\dfrac{d(\cot x)}{dx} = -\operatorname{cosec}^2 x$

· ·

Answer

6   (a) Let $u = \cos x$, so $\dfrac{du}{dx} = -\sin x$

Let $y = \ln u$, so $\dfrac{dy}{du} = \dfrac{1}{u} = \dfrac{1}{\cos x}$

$$\dfrac{dy}{dx} = \dfrac{du}{dx} \times \dfrac{dy}{du}$$

$$= (-\sin x)\left(\dfrac{1}{\cos x}\right)$$

$$= -\tan x$$

(b) Let $y = e^{6x}(3x - 2)^4$

Using the Product rule we have

$$\dfrac{dy}{dx} = 6e^{6x}(3x - 2)^4 + e^{6x} \times 4 \times 3(3x - 2)^3$$

$$= 6e^{6x}(3x - 2)^4 + 12e^{6x}(3x - 2)^3$$

$$= 6e^{6x}(3x - 2)^3(3x - 2 + 2)$$

$$= 18xe^{6x}(3x - 2)^3$$

(c) Let $y = \cot x = \dfrac{\cos x}{\sin x}$

$\cot x = \dfrac{1}{\tan x}$ and $\tan x = \dfrac{\sin x}{\cos x}$

so $\cot x = \dfrac{\cos x}{\sin x}$

Using the Quotient rule, we obtain

$$\dfrac{dy}{dx} = \dfrac{f'(x)\,g(x) - f(x)\,g'(x)}{(g(x))^2}$$

$$= \dfrac{(-\sin x)(\sin x) - (\cos x)(\cos x)}{\sin^2 x}$$

$$= \dfrac{-\sin^2 x - \cos^2 x}{\sin^2 x}$$

$$= \dfrac{-(\sin^2 x + \cos^2 x)}{\sin^2 x}$$

$$= \dfrac{-1}{\sin^2 x}$$

$$= -\operatorname{cosec}^2 x$$

Note that $\dfrac{1}{\sin^2 x} = \operatorname{cosec}^2 x$

5.10 Constructing simple differential equations

Differential equations often involve the rate at which a quantity changes with time.

For example, if the height of water in a tank is h metres, then $\frac{dh}{dt}$ is the rate at which the height varies with time.

If a spherical balloon is blown up, then the rate at which the radius, r, increases with time is $\frac{dr}{dt}$, the rate at which the surface area, A, increases is $\frac{dA}{dt}$ and the rate at which the volume, V, increases is $\frac{dV}{dt}$.

The surface area of a sphere is given by $A = 4\pi r^2$ so if you wanted to find the change in area with radius then this would be $\frac{dA}{dr}$. So to find $\frac{dA}{dr}$ you differentiate A with respect to r.

So, as $A = 4\pi r^2$, $\frac{dA}{dr} = 8\pi r$

If you knew the rate of change of radius was 0.5 cm s^{-1} you would find the rate of change of area when $r = 10$ cm using the following combinations of derivatives:

$$\frac{dA}{dt} = \frac{dA}{dr} \times \frac{dr}{dt} = 8\pi r \times 0.5 = 8\pi(10) \times 0.5 = 125.7 \text{ cm}^2 \text{ s}^{-1}$$

Examples

1   Air is pumped into a spherical balloon at the rate of 250 cm³ per second. When the radius of the balloon is 15 cm, calculate the rate at which the radius is increasing, giving your answer to three decimal places.   [3]

Answer

1   $\frac{dV}{dt} = 250$ cm³ s^{-1}

Now $V = \frac{4}{3}\pi r^3$ so $\frac{dV}{dr} = 3 \times \frac{4}{3}\pi r^2 = 4\pi r^2$

$\frac{dr}{dt} = \frac{dr}{dV} \times \frac{dV}{dt} = \frac{1}{4\pi r^2} \times 250 = \frac{1}{4\pi(15)^2} \times 250 = 0.088$ cm s^{-1} (3 d.p.)

---

2   A heap of cement is in the shape of a right circular cone. The cone grows in such as way that its height is always equal to the radius of the base.

If cement is added at the constant rate of 0.5 m³ per minute, find, to 2 significant figures, the rate at which the height is increasing when the height is 1.4 m.

Answer

2   Volume of a cone $= \frac{1}{3}\pi r^2 h$, but as $r = h$, we have $V = \frac{1}{3}\pi h^3$

$$\frac{dV}{dh} = \pi h^2$$

Now $$\frac{dV}{dt} = \frac{dV}{dh} \times \frac{dh}{dt}$$

$$0.5 = \pi h^2 \times \frac{dh}{dt}$$

When $h = 1.4$, we have, $0.5 = \pi(1.4)^2 \times \frac{dh}{dt}$,

giving $$\frac{dh}{dt} = 0.081 \text{ m per minute (2 s.f.)}$$

Test yourself

1 If $y = (4x^3 + 3x)^3$, find $\dfrac{dy}{dx}$. [2]

2 If $y = (3 - 2x)^{10}$, find $\dfrac{dy}{dx}$. [2]

3 A function is defined parametrically by $x = 3t^2$, $y = t^4$.
Find $\dfrac{dy}{dx}$ in terms of t. [2]

4 Find $\dfrac{dy}{dx}$ in terms of x and y for the curve $4x^3 - 6x^2 + 3xy = 5$. [3]

5 Differentiate each of the following with respect to x, simplifying your answer wherever possible.
(a) $\ln(x^3)$ [2]
(b) $\ln(\sin x)$ [2]

6 Differentiate each of the following with respect to x, simplifying your answer wherever possible.
(a) $(2x^2 - 1)^3$ [2]
(b) $x^3 \sin 2x$ [2]
(c) $\dfrac{3x^2 + 4}{x^2 + 6}$ [2]

7 Find $\dfrac{dy}{dx}$ for the equation $x^3 + 6xy^2 = y^3$. [3]

8 If $y = \ln(2 + 5x^2)$, find $\dfrac{dy}{dx}$. [3]

Summary

Check you know the following facts:

Finding points of inflection

To find points of inflection, find the second derivative and equate the resulting equation to zero. Solve the equation and put the *x*-value or values back into the equation of the curve to determine the *y*-coordinate(s). Check the sign of the second derivative either side of each *x*-coordinate to check that there is a sign change. If there is, then the point is a point of inflection.

Differentiation of e^{kx}, a^{kx}, $\sin kx$, $\cos kx$ and $\tan kx$

All these derivatives need to be remembered.

$$\frac{d(e^{kx})}{dx} = ke^{kx}$$

$$\frac{d(a^{kx})}{dx} = ka^{kx}\ln a$$

$$\frac{d(\sin kx)}{dx} = k\cos kx$$

$$\frac{d(\cos kx)}{dx} = -k\sin kx$$

The following derivative need not be remembered as it is included in the formula booklet.

$$\frac{d(\tan kx)}{dx} = k\sec^2 kx$$

The derivative of ln *x*

This derivative needs to be remembered.

$$\frac{d(\ln x)}{dx} = \frac{1}{x}$$

To differentiate the natural logarithm of a function you differentiate the function and then divide by the function.

This can be expressed mathematically as:

$$\frac{d(\ln(f(x)))}{dx} = \frac{f'(x)}{f(x)}$$

The Chain rule

If *y* is a function of *u* and *u* is a function of *x*, then the chain rule states:

$$\frac{dy}{dx} = \frac{dy}{du} \times \frac{du}{dx}$$

The Product rule

If $y = f(x)\,g(x)$, $\dfrac{dy}{dx} = f(x)\,g'(x) + g(x)\,f'(x)$

The Quotient rule

If $y = \dfrac{f(x)}{g(x)}$, $\dfrac{dy}{dx} = \dfrac{f'(x)\,g(x) - f(x)\,g'(x)}{(g(x))^2}$

Differentiation of simple functions defined implicitly

Finding $\dfrac{dy}{dx}$ in terms of both x and y is called implicit differentiation.

Here are the rules for differentiating implicitly:

- Terms involving x or constant terms are differentiated as normal.

- For terms just involving y, (e.g. $3y$, $5y^3$, etc.) differentiate with respect to y and then multiply the result by $\dfrac{dy}{dx}$.

- For terms involving both x and y (e.g. xy, $5x^2y^3$, etc.) the Product rule is used because there are two terms multiplied together. Note the need to include $\dfrac{dy}{dx}$ when the term involving y is differentiated.

Differentiation of functions defined parametrically

The equation of a curve can be expressed in parametric form by using:

$$x = f(t), y = g(t) \qquad \text{where } t \text{ is the parameter being used.}$$

The formulae for differentiating parametric forms are:

$$\frac{dy}{dx} = \frac{\dfrac{dy}{dt}}{\dfrac{dx}{dt}} = \frac{dy}{dt} \times \frac{dt}{dx}$$

and

$$\frac{d^2y}{dx^2} = \frac{\dfrac{d}{dt}\left(\dfrac{dy}{dx}\right)}{\dfrac{dx}{dt}} = \frac{d}{dt}\left(\frac{dy}{dx}\right) \times \frac{dt}{dx}$$

135

6 Coordinate geometry in the (x, y) plane

Introduction

This topic uses many of the mathematical techniques covered in previous topics. You must be proficient in the coordinate geometry material covered in Topic 3 of the AS book. You must also be proficient at basic differentiation, which was introduced in Topic 7 of the AS book and also in Topic 5 in this book.

This topic introduces you to a new way of representing the equation of the curve using equations called parametric equations.

This topic covers the following:

6.1 Parametric equations

6.2 Using the Chain rule to find the first derivative in terms of a parameter

6.3 Implicit differentiation

6.4 Using parametric equations in modelling in a variety of contexts

6.1 Parametric equations

Suppose we have two equations $x = 2 + t$ and $y = 5 + 3t$ where t can take any value.

Using $t = 0, 1, 2, 3, 4$, the following table is obtained:

t	0	1	2	3	4
x	2	3	4	5	6
y	5	8	11	14	17

You can see from the table that as x increases by 1, y increases by 3, meaning that the line has a gradient of 3.

As the point $(2, 5)$ lies on the line, the equation of the line is given by:

$$y - y_1 = m(x - x_1)$$

So $\qquad y - 5 = 3(x - 2)$ giving $y = 3x - 1$

Hence the equation of the straight line is $y = 3x - 1$.

This equation connects x and y and is called the Cartesian equation. The original equations (i.e. $x = 2 + t$ and $y = 5 + 3t$) are called the parametric equations.

If you are given two parametric equations such as $x = 2 + t$ and $y = 5 + 3t$ and want to find the Cartesian equation, you can do this by eliminating the parameter, t.

$$x = 2 + t \text{ so } t = x - 2$$

$$y = 5 + 3t \text{ so } y = 5 + 3(x - 2)$$

Hence $\qquad y = 5 + 3x - 6$

giving the Cartesian equation $y = 3x - 1$.

Parametric equations can also represent curves. For example, the parametric equations $x = 2t^2$, $y = 4t$ represent a parabola.

As with the parametric equations of a straight line, to find the Cartesian equation of a curve, the parameter t must be eliminated leaving the equation in terms of x and y.

Suppose you are given the parametric equations $x = 2t^2$, $y = 4t$ and have to find the Cartesian equation of the curve.

From $y = 4t$, we have $t = \dfrac{y}{4}$

Substituting this for t into $x = 2t^2$ gives $x = 2\left(\dfrac{y}{4}\right)^2$

So $\qquad\qquad\qquad\qquad x = \dfrac{y^2}{8}$ or $y^2 = 8x$

This is now the Cartesian equation as it connects x and y.

Examples

1   Find the Cartesian equation of the line given by the parametric equations

$$x = 4 + 2t \qquad\qquad y = 1 + 2t$$

Note that to find the Cartesian equation, the parameter t is eliminated.

Answer

1 $x = 4 + 2t$ so $2t = x - 4$

Substituting $2t = x - 4$ for $2t$ in the equation $y = 1 + 2t$ gives

$$y = 1 + x - 4$$

$$y = x - 3$$

---

2   A curve is given by the parametric equations $x = t^2, y = t^3$.

Find the equation of the tangent to the curve at the point given by $t = 2$.

Note that

$$\frac{dy}{dx} = \frac{1}{\frac{dx}{dt}}$$

so you must remember to invert $\frac{dx}{dt}$.

Answer

2 $\dfrac{dx}{dt} = 2t$ and $\dfrac{dy}{dt} = 3t^2$

We use the Chain rule to find $\dfrac{dy}{dx}$

Hence $\dfrac{dy}{dx} = \dfrac{dy}{dt} \times \dfrac{dt}{dx} = 3t^2 \times \dfrac{1}{2t} = \dfrac{3}{2}t$

When $t = 2$, $\dfrac{dy}{dx} = \dfrac{3}{2} \times 2 = 3$

$t = 2$ is substituted for t in the parametric equations $x = t^2$ and $y = t^3$.

When $t = 2$, $x = 2^2 = 4$ and $y = 2^3 = 8$

Hence point $(4, 8)$ lies on the curve.

Equation of tangent passing through $(4, 8)$ and having gradient 3 is

$$y - 8 = 3(x - 4)$$

Use the formula for a straight line $y - y_1 = m(x - x_1)$.

Hence $y - 3x + 4 = 0$

---

3   The curve C has the parametric equations

$$x = 3 \cos t, y = 4 \sin t.$$

The point P lies on C and has parameter p.

(a)   Show that the equation of the tangent to C at the point P is

$$(3 \sin p)y + (4 \cos p)x - 12 = 0$$

(b)   The tangent to C at the point P meets the x-axis at the point A and the y-axis at the point B. Given that

$$p = \frac{\pi}{6}$$

(i)   find the coordinates of A and B,

(ii)   show that the exact length of AB is $2\sqrt{19}$.

Answer

3   (a) $x = 3 \cos t$, $y = 4 \sin t$

Differentiating both with respect to t gives:

$$\frac{dx}{dt} = -3 \sin t, \quad \frac{dy}{dt} = 4 \cos t$$

$$\frac{dy}{dx} = \frac{dy}{dt} \times \frac{dt}{dx}$$

$$= 4\cos t \times \frac{1}{-3\sin t}$$

$$= -\frac{4\cos t}{3\sin t}$$

> Use $\dfrac{dt}{dx} = \dfrac{1}{\frac{dx}{dt}}$

At P($3\cos p$, $4\sin p$) the gradient $= -\dfrac{4\cos p}{3\sin p}$

> Parameter p indicates that $x = 3\cos p$, $y = 4\sin p$.

Equation of the tangent to C at P is given by

$$y - y_1 = m(x - x_1)$$

$$y - 4\sin p = -\frac{4\cos p}{3\sin p}\left(x - 3\cos p\right)$$

$$3y\sin p + 4x\cos p = 12\,(\sin^2 p + \cos^2 p)$$

> Use $\sin^2 p + \cos^2 p = 1$

Hence $\quad 3y\sin p + 4x\cos p = 12$

So $(3\sin p)y + (4\cos p)x - 12 = 0$

(b) (i) Now $p = \dfrac{\pi}{6}$ and at A, $y = 0$

> $\sin\dfrac{\pi}{6} = \dfrac{1}{2}$ and $\cos\dfrac{\pi}{6} = \dfrac{\sqrt{3}}{2}$.

Hence $\left(3\sin\dfrac{\pi}{6}\right)(0) + \left(4\cos\dfrac{\pi}{6}\right)x - 12 = 0$

$$0 + \frac{4\sqrt{3}}{2}x - 12 = 0$$

$$2\sqrt{3}x = 12$$

$$x = \frac{6}{\sqrt{3}}$$

$$= \frac{6\sqrt{3}}{\sqrt{3}\,\sqrt{3}} = 2\sqrt{3}$$

> Remember to rationalise the denominator by multiplying the top and bottom by $\sqrt{3}$.

Point A has coordinates $(2\sqrt{3}, 0)$

For point B, $x = 0$.

Hence $\quad \left(3\sin\dfrac{\pi}{6}\right)y + 0 - 12 = 0$

$$\left(\sin\dfrac{\pi}{6}\right)y = 4$$

$$\frac{1}{2}y = 4$$

$$y = 8$$

Point B has coordinates $(0, 8)$

(ii)   The length of a straight line joining the two points (x_1, y_1) and (x_2, y_2) is given by:

$$\sqrt{(x_2 - x_1)^2 + (y_2 - y_1)^2}$$

Hence distance between A $(2\sqrt{3}, 0)$ and B $(0, 8)$ is:

$$= \sqrt{(2\sqrt{3} - 0)^2 + (0 - 8)^2}$$
$$= \sqrt{12 + 64}$$
$$= \sqrt{76}$$
$$= \sqrt{4 \times 19}$$
$$= 2\sqrt{19}$$

6.2 Using the Chain rule to find the first derivative in terms of a parameter

The Chain rule can be used to find the first derivative in terms of a parameter such as t in the following way:

$$\frac{dy}{dx} = \frac{dy}{dt} \times \frac{dt}{dx}$$

The following example shows this technique.

Example

1   The curve C has the parametric equations $x = t^2$, $y = t^2 + 2t$.

Find $\dfrac{dy}{dx}$ in terms of t, simplifying your result as much as possible.

· ·

Answer

1        $x = t^2$, $y = t^2 + 2t$

$$\frac{dx}{dt} = 2t \qquad \frac{dy}{dt} = 2t + 2$$

$$\frac{dy}{dx} = \frac{dy}{dt} \times \frac{dt}{dx} = (2t + 2) \times \frac{1}{2t} = \frac{t + 1}{t} = 1 + \frac{1}{t}$$

6.3 Implicit differentiation

Implicit differentiation was covered in Topic 5 of this book.

To recap, finding $\frac{dy}{dx}$ in terms of both x and y is called implicit differentiation. There are a number of rules and these are summarised here.

Here are the rules for differentiating implicitly:

- Terms involving x or constant terms are differentiated as normal.

- For those terms just involving y, (e.g. $3y$, $5y^3$, etc.) differentiate with respect to y and then multiply the result by $\frac{dy}{dx}$.

- For terms involving both x and y (e.g. xy, $5x^2y^3$, etc.) the Product rule is used because there are two terms multiplied together. Note the need to include $\frac{dy}{dx}$ when the term involving y is differentiated.

The following examples should be read after briefly looking over Topic 5 of this book.

Examples

1 Find the equation of the tangent to the curve

$$6x^2 + xy + 3y^2 = 6$$

at point $(1, 0)$.

. . . .

Answer

1   $6x^2 + xy + 3y^2 = 6$

Differentiating with respect to x gives

$$12x + (x)\frac{dy}{dx} + y(1) + 6y\frac{dy}{dx} = 0$$

$$\frac{dy}{dx}(x + 6y) = -12x - y$$

Hence

$$\frac{dy}{dx} = \frac{-12x - y}{x + 6y}$$

At $(1, 0)$,

$$\frac{dy}{dx} = \frac{-12 - 0}{1 + 0} = -12$$

Equation of the tangent at the point $(1, 0)$ is

$$y - 0 = -12(x - 1)$$

$$y = -12x + 12$$

> When differentiating $f(y)$ with respect to x, we obtain
> $$f'(y)\frac{dy}{dx}$$

> The term xy is differentiated using the Product rule.

> $\frac{dy}{dx}$ is taken out as a factor.

> Use $y - y_1 = m(x - x_1)$ to find the equation of the tangent.

2 Find the equation of the normal to the curve

$$5x^2 + 4xy - y^3 = 5$$

at point $(1, -2)$.

Answer

2 $5x^2 + 4xy - y^3 = 5$

Differentiating with respect to x gives

The Product rule is used to differentiate $4xy$.

$$10x + (4x)\frac{dy}{dx} + y(4) - 3y^2\frac{dy}{dx} = 0$$

$$10x + 4y = \frac{dy}{dx}(3y^2 - 4x)$$

Hence

$$\frac{dy}{dx} = \frac{10x + 4y}{3y^2 - 4x}$$

The normal and the tangent at the same point are at right angles to each other, so the product of their gradients is −1.

At $(1, -2)$, $\frac{dy}{dx} = \frac{10 - 8}{12 - 4} = \frac{1}{4}$

Gradient of the normal = −4

Equation of the normal at the point $(1, -2)$ is

Use $y - y_1 = m(x - x_1)$ to find the equation of the normal.

$$y + 2 = -4(x - 1)$$

$$y = -4x + 2$$

---

3 The parametric equations of the curve C are

$$x = \frac{2}{t}, \quad y = 4t$$

(a) Show that the tangent to C at the point P with parameter p has equation

$$y = -2p^2x + 8p$$

(b) The tangent to C at the point P passes through the point $(2, 3)$. Show that P can be one of two points.

Find the coordinates of each of these two points.

Answer

3 (a) $x = \frac{2}{t}, \quad y = 4t$

$x = 2t^{-1}$ so $\frac{dx}{dt} = -2t^{-2} = \frac{-2}{t^2}$

$$\frac{dy}{dt} = 4$$

$$\frac{dy}{dx} = \frac{dy}{dt} \times \frac{dt}{dx} = 4 \times \frac{t^2}{-2} = -2t^2$$

Change the parameter from t to p.

At P, $x = \frac{2}{p}, \quad y = 4p$

Equation of the tangent at P is $y - 4p = -2p^2\left(x - \frac{2}{p}\right)$

Use $y - y_1 = m(x - x_1)$.

$$y - 4p = -2p^2x + 4p$$

$$y = -2p^2x + 8p$$

(b) The coordinates (2, 3) must satisfy the equation of the tangent. Hence

$$y = -2p^2x + 8p$$

so

$$3 = -4p^2 + 8p$$

$$4p^2 - 8p + 3 = 0$$

$$(2p - 1)(2p - 3) = 0$$

giving $p = \dfrac{1}{2}, \dfrac{3}{2}$.

Hence the coordinates of the two points are $(4, 2)$, $\left(\dfrac{4}{3}, 6\right)$

Substitute each value of the parameter p giving

$$x = \frac{2}{p}, y = 4p$$

to find both pairs of coordinates.

---

4 (a) The curve C is given by the equation

$$x^4 + x^2y + y^2 = 13$$

Find the value of $\dfrac{dy}{dx}$ at the point $(-1, 3)$.

(b) Show that the equation of the normal to the curve $y^2 = 4x$ at the point P $(p^2, 2p)$ is:

$$y + px = 2p + p^3$$

Given that $p \neq 0$ and that the normal at P cuts the x-axis at B $(b, 0)$, show that $b > 2$.

. . . .

Answer

4 (a) $x^4 + x^2y + y^2 = 13$

Differentiating with respect to x gives

$$4x^3 + x^2\frac{dy}{dx} + y(2x) + 2y\frac{dy}{dx} = 0$$

As $x = -1$ and $y = 3$, we have:

$$4(-1)^3 + (-1)^2\frac{dy}{dx} + 3(2(-1)) + 2(3)\frac{dy}{dx} = 0$$

$$-4 + \frac{dy}{dx} - 6 + 6\frac{dy}{dx} = 0$$

$$7\frac{dy}{dx} = 10$$

$$\frac{dy}{dx} = \frac{10}{7}$$

(b) $y^2 = 4x$

Differentiating with respect to x gives $\quad 2y\dfrac{dy}{dx} = 4$

Hence, $\qquad\qquad\qquad\qquad\qquad \dfrac{dy}{dx} = \dfrac{2}{y}$

Now at P $(p^2, 2p)$ gradient of tangent $= \dfrac{2}{2p} = \dfrac{1}{p}$

So gradient of normal $= -p$

BOOST

Grade ⇧⇧⇧⇧

Make sure that you differentiate the right-hand side of the equation as well as the left.

Here we use $m_1 m_2 = -1$

Equation of normal is given by:

$$y - 2p = -p(x - p^2)$$

$$y - 2p = -px + p^3$$

$$y + px = 2p + p^3$$

Equation of the x-axis is $y = 0$

Hence $0 + px = 2p + p^3$

So, $x = 2 + p^2$

At B, $x = b$, so $b = 2 + p^2$ and $b - 2 = p^2$

For $p^2 > 0$, $b > 2$

$b - 2$ must be positive.

Step by STEP

The curve C is defined by

$$y^4 - 2x^2 + 8xy^2 + 9 = 0$$

Show that there is no point on C at which $\dfrac{dy}{dx} = 0$.

Steps to take

1   We need to find an expression for the gradient so we start by differentiating implicitly with respect to x.

2   When differentiating, the Product rule needs to be used to differentiate $8xy^2$.

3   Rearrange the equation to make $\dfrac{dy}{dx}$ the subject.

4   Show that the resulting equation cannot have a value of zero.

· ·

Answer

$$y^4 - 2x^2 + 8xy^2 + 9 = 0$$

Differentiating implicitly with respect to x.

$$4y^3\frac{dy}{dx} - 4x + 8x(2y)\frac{dy}{dx} + y^2(8) = 0$$

$$4y^3\frac{dy}{dx} - 4x + 16xy\frac{dy}{dx} + 8y^2 = 0$$

$$4y^3\frac{dy}{dx} + 16xy\frac{dy}{dx} = 4x - 8y^2$$

$$\frac{dy}{dx}(4y^3 + 16xy) = 4(x - 2y^2)$$

$$\frac{dy}{dx} = \frac{4(x - 2y^2)}{4y(y^2 + 4x)}$$

$$= \frac{x - 2y^2}{y(y^2 + 4x)}$$

For $\dfrac{dy}{dx} = 0$, the numerator must be zero.

Hence, $x - 2y^2 = 0$, so $x = 2y^2$

Substituting $x = 2y^2$ into the equation of C, we obtain:

$$y^4 - 2(2y^2)^2 + 8(2y^2)y^2 + 9 = 0$$

$$y^4 - 8y^4 + 16y^4 + 9 = 0$$

$$9y^4 + 9 = 0$$

$$y^4 = -1 \text{ (which is impossible)}$$

Hence no real point exists where $\dfrac{dy}{dx} = 0$.

> y^4 will always be positive for positive and negative values of y.

6.4 Using parametric equations in modelling in a variety of contexts

Parametric equations can be used to model situations, particularly in mechanics.

Step by STEP

A spider and a fly are on the (x, y) plane.

The spider's position at time t seconds is given by $x = 10t$, $y = t^2$.

The fly's position at time t seconds is given by $x = 2t$, $y = t - 6$.

(a) Determine algebraically whether the paths taken by each insect cross.

(b) Show whether the spider and fly will meet.

Steps to take

1   Notice the x- and y-positions of the insects are given in terms of the parameter t and that t represents time in seconds.

2   Find the Cartesian equations of each path by eliminating the parameter t. The resulting equation will only be in terms of x and y.

3   Solve the two Cartesian equations of the paths simultaneously to find the coordinates of any points of intersection.

4   Use the x- or y-values to find the value of t for each point of intersection, for each insect. This will determine if they are at the same position at the same time which means they would meet.

Answer

$x = 10t$, $y = t^2$

$x^2 = 100t^2$, so substituting in for y we obtain:

$$x^2 = 100y$$

$x = 2t$, $y = t - 6$

$t = \dfrac{x}{2}$,   so $y = \dfrac{x}{2} - 6$

> This is the Cartesian equation for the path of the spider.

This is the Cartesian equation for the path of the fly.

$$x = 2y + 12$$

Solving the two Cartesian equations simultaneously we have:

$$x^2 = (2y + 12)^2, \text{ but as } x^2 = 100y \text{ we have:}$$

$$100y = (2y + 12)^2$$

$$100y = 4y^2 + 48y + 144$$

$$4y^2 - 52y + 144 = 0$$

$$y^2 - 13y + 36 = 0$$

$$(y - 9)(y - 4) = 0$$

Hence $y = 9$ or 4 and substituting these values into $x = 2y + 12$ gives $x = 30$ or 20.

Coordinates of points where paths intersect are $(30, 9)$ and $(20, 4)$.

We now need to find the times when each insect are at these coordinates.

First we will look at the point $(30, 9)$

Note that these coordinates are only where the paths intersect. For the insects to meet they would have to be at a particular coordinate at the same time.

For the spider, $x = 10t$, $y = t^2$

When $x = 30$, $t = 3$ s

For the fly, $x = 2t$, $y = t - 6$

When $x = 30$, $t = 15$ s

Hence the spider and fly will not meet as they are not at this point at the same time.

Now we will look at the point $(20, 4)$

For the spider $x = 10t$, $y = t^2$

When $x = 20$, $t = 2$ s

For the fly $x = 2t$, $y = t - 6$

When $x = 20$, $t = 10$ s

Hence the spider and fly will not meet as they are not at this point at the same time.

So as the paths taken by the insects intersect twice, the insects are not at either point at the same time so they will not meet.

Test yourself

1 The parametric equations of the curve C are $x = 3t^2, y = t^3$.
The point P has parameter p.
Show that the equation of the normal to C at the point P is
$py + 2x = p^2(6 + p^2)$. [5]

2 Given that $y^2 - 5xy + 8x^2 = 2$, prove that $\dfrac{dy}{dx} = \dfrac{5y - 16x}{2y - 5x}$ [4]

3 The parametric equations of the curve C are $x = 2\sin 4t, \ y = \cos 4t$.

(a) Prove that $\dfrac{dy}{dx} = -\dfrac{1}{2}\tan 4t$. [3]

(b) Show that the equation of the tangent to C at the point P with
parameter p is
$$2y\cos 4p + x\sin 4p = 2$$ [4]

4 The parametric equations of the curve C are $x = t^2, y = t^3$. The point P has
parameter p.

(a) Show that the gradient of the tangent to C at the point P is $\dfrac{3}{2}p$. [4]

(b) Find the equation of the tangent at P. [2]

5 The curve C has equation $4x^2 - 6xy + y^2 = 20$

(a) Prove that $\dfrac{dy}{dx} = \dfrac{3y - 4x}{y - 3x}$. [4]

(b) Points A and B lie on C. If the x-coordinates of A and B are both equal
to 0, prove that the y-coordinates of A and B are $\pm 2\sqrt{5}$. [3]

6 A toy car A travels along a path given by $x = 40t - 40$ and $y = 120t - 160$
where t is the time in seconds and the distances are in metres. A toy car B
travels in the same horizontal plane and its path is given by $x = 30t, y = 20t^2$.
(a) Find the coordinates where the paths meet. [3]
(b) Determine whether or not the toy cars collide. [3]

Summary

Check you know the following facts:

Cartesian and parametric equations

Cartesian equations connect x and y in some way. For example, $y = 4x^3$ is a Cartesian equation.

Parametric equations express x and y in terms of a parameter such as t, for example

$$x = 4 + 2t \qquad y = 1 + 2t$$

To obtain the Cartesian equation from the parametric equation it is necessary to eliminate the parameter.

$$\text{Note that} \quad \frac{dy}{dx} = \frac{dy}{dt} \times \frac{dt}{dx}$$

Using the Chain rule to find the second derivative

The Chain rule can be used to find the second derivative in terms of a parameter such as t in the following way:

$$\frac{d^2y}{dx^2} = \frac{d}{dx}\left(\frac{dy}{dx}\right) = \frac{d}{dt}\left(\frac{dy}{dx}\right)\frac{dt}{dx}$$

Implicit differentiation

Here are the basic rules:

$$\frac{d(3x^2)}{dx} = 6x$$

> Terms involving x or constant terms are differentiated as normal.

$$\frac{d(6y^3)}{dx} = 18y^2 \times \frac{dy}{dx}$$

> Differentiate with respect to y and then multiply the result by $\frac{dy}{dx}$.

$$\frac{d(y)}{dx} = 1 \times \frac{dy}{dx}$$

> When you are differentiating a term just involving y, you differentiate with respect to y and then multiply the result by $\frac{dy}{dx}$. This is an application of the Chain rule.

$$\frac{d(x^2y^3)}{dx} = (x^2)\left(3y^2 \times \frac{dy}{dx}\right) + (y^3)(2x)$$

$$= 3x^2y^2\frac{dy}{dx} + 2xy^3$$

> Because there are two terms here, the Product rule is used. Notice the need to include $\frac{dy}{dx}$ when the term involving y is differentiated.

7 Integration

Introduction

You came across integration in Topic 8 of the AS Pure 1 course. Integration is the opposite of differentiation, so to integrate a term, you add one to the index and then divide by the new index. With indefinite integration you must remember to include a constant of integration and with definite integration you substitute in numbers. Definite integrals represent the area under the curve. As you will be building on the material from Topic 8 of the AS Pure course, you should look over that material before starting this topic.

This topic covers the following:

7.1 Integration of $x^n (n \neq -1)$, e^{kx}, $\frac{1}{x}$, $\sin kx$, $\cos kx$

7.2 Integration of $(ax + b)^n (n \neq -1)$, e^{ax+b}, $\frac{1}{ax + b}$, $\sin(ax + b)$, $\cos(ax + b)$

7.3 Using definite integration to find the area between two curves

7.4 Using integration as the limit of a sum

7.5 Integration by substitution and integration by parts

7.6 Integration using partial fractions

7.7 Analytical solution of first order differential equations with separable variables

7.1 Integration of $x^n (n \neq -1)$, e^{kx}, $\frac{1}{x}$, $\sin kx$, $\cos kx$

Integration is the reverse process to differentiation and this fact can be used to help remember the integrals of the functions covered in this section as the formulas for these integrals are not included in the formula booklet. Always remember, if you integrate without using limits you must include a constant of integration, c.

Integration of $x^n (n \neq -1)$

$$\int x^n \, dx = \frac{x^{n+1}}{n+1} + c$$

Integration of e^{kx}

$$\int e^{kx} \, dx = \frac{e^{kx}}{k} + c$$

Integration of $\frac{1}{x}$

You cannot find the ln of a negative number so the modulus sign is included here.

$$\int \frac{1}{x} \, dx = \ln |x| + c$$

Integration of $\sin kx$

Note that the derivative of $\cos x$ is $-\sin x$, so that the integral of $\sin x$ is $-\cos x$, as integration is the opposite process to differentiation.

$$\int \sin kx \, dx = -\frac{1}{k} \cos kx + c$$

Integration of $\cos kx$

If you differentiate $\sin x$ you obtain $\cos x$.

$$\int \cos kx \, dx = \frac{1}{k} \sin kx + c$$

Examples

1 Find $\int (4x^3 + 3x^2 + 2x + 1) dx$

Answer

1 $\int (4x^3 + 3x^2 + 2x + 1) dx = \dfrac{4x^4}{4} + \dfrac{3x^3}{3} + \dfrac{2x^2}{2} + x + c$

$$= x^4 + x^3 + x^2 + x + c$$

2 Find $\int \dfrac{1}{2x} \, dx$

Answer

2 $\int \dfrac{1}{2x} \, dx = \dfrac{1}{2} \int \dfrac{1}{x} \, dx$

$$= \frac{1}{2} \ln |x| + c$$

3 Find $\int e^{3x} \, dx$

Answer

3 $\int e^{3x} \, dx = \dfrac{1}{3} e^{3x} + c$

4 Find $\int 3\cos 3x\, dx$

. . . .

Answer

4 $\quad \int 3\cos 3x\, dx = 3 \int \cos 3x\, dx$

$\qquad\qquad = 3 \times \dfrac{1}{3} \sin x + c$

$\qquad\qquad = \sin x + c$

---

5 Find $\int \dfrac{1}{2} \sin \dfrac{x}{2}\, dx$

. . . .

Answer

5 $\quad \int \dfrac{1}{2} \sin \dfrac{x}{2}\, dx = \dfrac{1}{2} \int \sin \dfrac{x}{2}\, dx$

$\qquad\qquad = -\cos \dfrac{x}{2} + c$

7.2 Integration of $(ax + b)^n$ $(n \neq -1)$, e^{ax+b}, $\dfrac{1}{ax+b}$, $\sin(ax + b)$, $\cos(ax + b)$

The results in the table earlier on are modified when x is replaced by the linear expression $ax + b$, where a and b are constants. Thus, for example,

when, $n \neq -1$, $\quad \dfrac{d}{dx}\left((ax + b)^{n+1}\right) = (n + 1)(ax + b)^n (a)$

> Using the Chain rule.

so that $\quad \int (n + 1)(ax + b)^n (a)\, dx = (ax + b)^{n+1}$

> Omitting the constant of integration.

or on division of both sides by the constant $(n + 1)a$,

$$\int (ax + b)^n\, dx = \frac{(ax + b)^{n+1}}{(n + 1)a} + c$$

Similarly, since

$$\frac{d}{dx}\left(e^{ax+b}\right) = \left(e^{ax+b}\right)(a) + c$$

we have $\quad \displaystyle\int e^{ax+b}\, dx = \frac{e^{ax+b}}{a} + c$

> Using the Chain rule and omitting the constant of integration.

The full table is then as follows:

$$\int (ax + b)^n\, dx = \frac{(ax + b)^{n+1}}{(n + 1)a} + c \qquad (n \neq -1)$$

$$\int e^{ax+b}\, dx = \frac{e^{ax+b}}{a} + c$$

> You will be required to remember these results as they are not given in the formula booklet.

$$\int \frac{1}{ax + b}\, dx = \frac{1}{a} \ln |ax + b| + c$$

$$\int \sin(ax + b)\, dx = \frac{-\cos(ax + b)}{a} + c$$

$$\int \cos(ax + b)\, dx = \frac{\sin(ax + b)}{a} + c$$

First table is given on page 150.

If you differentiate
$$\frac{(ax^2 + b)^{n+1}}{(n+1)2ax}$$
you will not obtain $(ax^2 + b)^n$.
Remember the Quotient rule.

It is appropriate to draw attention to two aspects of the modification of the first table to the second table.

Firstly, the results in the second table are obtained from those of the first by:

- replacing x by $ax + b$,
- introducing a factor of $\frac{1}{a}$.

Secondly, it is important to notice that the relatively simple modification occurs because the derivative of $ax + b$ is a.

Thus note that

$$\int (ax^2 + b)^n \, dx \text{ is } \textbf{not} \text{ equal to } \frac{(ax^2 + b)^{n+1}}{(n+1)2ax} + c$$

Examples

1 Where possible, use the second table to integrate the following. If not, explain why you cannot use the second table.

(a) $\dfrac{1}{4x + 5}$

(b) $\sin(x^3)$

(c) $\cos 3x$

(d) $\dfrac{1}{5x^2 + 7}$

· ·

Answers

1   (a) $4x + 5$ is of the form $ax + b$, where $a = 4$ and $b = 5$.

Then $\displaystyle\int \frac{1}{4x + 5} \, dx = \frac{1}{4} \ln |4x + 5| + c$

(b) The table cannot be used to find $\int \sin(x^3)$ because x^3 is not of the form $ax + b$.

(c) $3x$ is of the form $ax + b$, where $a = 3$ and $b = 0$.

Then $\displaystyle\int \cos 3x \, dx = \frac{\sin 3x}{3} + c$

(d) $5x^2 + 7$ is not of the form $ax + b$, so the table cannot be used.

---

2 Integrate

(a) e^{2x}

(b) $\sin(7x + 5)$

(c) $\dfrac{1}{7x + 1}$

(d) $\cos\left(\dfrac{x}{3}\right)$

Answers

2   (a) $\dfrac{e^{2x}}{2} + c$

e^{ax+b} with $a = 2$, $b = 0$.

(b) $\dfrac{-\cos(7x + 5)}{7} + c$

$\sin(ax + b)$ with $a = 7$, $b = 5$.

(c) $\dfrac{1}{7}\ln|7x + 1| + c$

$\dfrac{1}{ax+b}$ with $a = 7$, $b = 1$.

(d) $\dfrac{\sin\left(\frac{x}{3}\right)}{\left(\frac{1}{3}\right)} + c = 3\sin\left(\dfrac{x}{3}\right) + c$

$\cos(ax + b)$ with $a = \dfrac{1}{3}$, $b = 0$.

We note also that $\int k(ax + b)\, dx = k \int (ax + b)\, dx$ and that definite integrals are evaluated as in the AS course.

Examples

1 Integrate

(a) $6e^{-2x}$

(b) $\dfrac{8}{4x + 1}$

(c) $7\sin(2x + 3)$

(d) $15\cos(3x + 2)$

Answers

1   (a) $\int 6e^{-2x}\, dx = 6\int e^{-2x}\, dx$

e^{ax+b} with $a = -2$, $b = 0$.

$= 6 \times \dfrac{e^{-2x}}{-2} + c$

$= -3e^{-2x} + c$

(b) $\int \dfrac{8}{4x + 1}\, dx = 8 \int \dfrac{1}{4x + 1}\, dx$

$\dfrac{1}{ax+b}$ with $a = 4$, $b = 1$.

$= 8 \times \dfrac{1}{4}\ln|4x + 1| + c$

$= 2\ln|4x + 1| + c$

(c) $\int 7\sin(2x + 3)\, dx = 7\int \sin(2x + 3) + c$

$\sin(ax + b)$ with $a = 2$, $b = 3$.

$= -\dfrac{7}{2}\cos(2x + 3) + c$

(d) $\int 15\cos(3x + 2)\, dx = 15\int \cos(3x + 2) + c$

$= 15 \times \dfrac{1}{3}\sin(3x + 2) + c$

$\cos(ax + b)$ with $a = 3$, $b = 2$.

$= 5\sin(3x + 2) + c$

2   Find the values of

(a) $\displaystyle\int_{2}^{4} \frac{8}{(3x-4)^3}\, dx$

(b) $\displaystyle\int_{0}^{\frac{\pi}{6}} 3 \sin\left(4x + \frac{\pi}{6}\right) dx$

• •

Answers

$(ax + b)^n$ with $a = 3$, $b = -4$ and $n = -3$.

2   (a) $\displaystyle\int_{2}^{4} \frac{8}{(3x-4)^3}\, dx = 8\int_{2}^{4} (3x-4)^{-3}\, dx$

$$= 8\left[\frac{(3x-4)^{-2}}{(-2) \times 3}\right]_{2}^{4}$$

$$= -\frac{8}{6}\left[(3x-4)^{-2}\right]_{2}^{4}$$

$$= -\frac{4}{3}\left[(12-4)^{-2} - (6-4)^{-2}\right]$$

$$= -\frac{4}{3}\left[\frac{1}{64} - \frac{1}{4}\right]$$

$$= -\frac{4}{3} \times \left(-\frac{15}{64}\right)$$

$$= \frac{5}{16}$$

$\sin(ax + b)$ with $a = 4$, $b = \frac{\pi}{6}$.

(b) $\displaystyle\int_{0}^{\frac{\pi}{6}} 3 \sin\left(4x + \frac{\pi}{6}\right) dx = 3\int_{0}^{\frac{\pi}{6}} \sin\left(4x + \frac{\pi}{6}\right) dx$

$$= 3\left[-\frac{1}{4}\cos\left(4x + \frac{\pi}{6}\right)\right]_{0}^{\frac{\pi}{6}}$$

$$= 3\left[-\frac{1}{4}\cos\frac{5\pi}{6} + \frac{1}{4}\cos\frac{\pi}{6}\right]$$

$\cos\frac{5\pi}{6} = -\cos\frac{\pi}{6}.$

$$= 3\left[\frac{1}{4}\cos\frac{\pi}{6} + \frac{1}{4}\cos\frac{\pi}{6}\right]$$

$$= 3 \times \frac{1}{2}\cos\frac{\pi}{6}$$

$$= 3 \times \frac{1}{2} \times \frac{\sqrt{3}}{2}$$

$$= \frac{3\sqrt{3}}{4}$$

$$= 1.299 \text{ (correct to three decimal places)}.$$

3   Evaluate $\displaystyle\int_{0}^{\frac{\pi}{3}} \cos\left(6x + \frac{\pi}{3}\right) dx$

Answer

3 $\displaystyle\int_0^{\frac{\pi}{3}} \cos\left(6x + \frac{\pi}{3}\right) dx = \left[\frac{1}{6}\sin\left(6x + \frac{\pi}{3}\right)\right]_0^{\frac{\pi}{3}}$

$$= \frac{1}{6}\left[\sin\left(6x + \frac{\pi}{3}\right)\right]_0^{\frac{\pi}{3}}$$

$$= \frac{1}{6}\left[\left(\sin\frac{7\pi}{3}\right) - \left(\sin\frac{\pi}{3}\right)\right]$$

$$= \frac{1}{6}\left[\frac{\sqrt{3}}{2} - \frac{\sqrt{3}}{2}\right]$$

$$= 0$$

$\cos(ax + b)$ with $a = 6$, $b = \frac{\pi}{3}$.

---

4 Find

(a) $\displaystyle\int \frac{1}{1 - 2x}\, dx$

(b) $\displaystyle\int 6e^{6x}\, dx$

(c) $\displaystyle\int (4x - 3)^4\, dx$

Answer

4 (a) $\displaystyle\int \frac{1}{1 - 2x}\, dx = -\frac{1}{2}\ln|1 - 2x| + c$

$\frac{1}{ax + b}$ with $a = -2$, $b = 1$.

(b) $\displaystyle\int 6e^{6x}\, dx = \frac{6}{6}e^{6x} + c = e^{6x} + c$

e^{ax+b} with $a = 6$, $b = 0$.

(c) $\displaystyle\int (4x - 3)^4\, dx = \frac{(4x - 3)^5}{4 \times 5} + c$

$$= \frac{1}{20}(4x - 3)^5 + c$$

$(ax + b)^n$ with $a = 4$, $b = -3$ and $n = 4$.

---

5 (a) Find

(i) $\displaystyle\int \cos 4x\, dx$

(ii) $\displaystyle\int 5e^{2-3x}\, dx$

(iii) $\displaystyle\int \frac{3}{(6x - 7)^5}\, dx$

(b) Evaluate $\displaystyle\int_1^4 \frac{9}{2x + 5}\, dx$, giving your answer correct to three decimal places.

Answer

5 (a) (i) $\displaystyle\int \cos 4x\, dx = \frac{1}{4}\sin 4x + c$

(ii) $\int 5e^{2-3x}\,dx = \dfrac{5}{-3}e^{2-3x} + c$

$\qquad\qquad\qquad = -\dfrac{5}{3}e^{2-3x} + c$

(iii) $\int \dfrac{3}{(6x-7)^5}\,dx = 3\int (6x-7)^{-5}\,dx$

$\qquad\qquad\qquad\qquad = \dfrac{3}{-4\times 6}(6x-7)^{-4} + c$

$\qquad\qquad\qquad\qquad = -\dfrac{1}{8}(6x-7)^{-4} + c$

(b) $\displaystyle\int_1^4 \dfrac{9}{2x+5}\,dx = \dfrac{9}{2}\Big[\ln |2x+5|\Big]_1^4$

$\qquad\qquad\qquad = \dfrac{9}{2}\Big[\ln 13 - \ln 7\Big]$

$\qquad\qquad\qquad = \dfrac{9}{2}\ln\dfrac{13}{7}$

$\qquad\qquad\qquad = 2.786$ (correct to three decimal places)

> Use one of the three laws of logarithms:
> $$\log_a x - \log_a y = \log_a \dfrac{x}{y}$$

---

6 (a) Find

(i) $\displaystyle\int \dfrac{1}{4x-7}\,dx$

(ii) $\displaystyle\int e^{3x-1}\,dx$

(iii) $\displaystyle\int \dfrac{5}{(2x+3)^4}\,dx$

(b) Evaluate $\displaystyle\int_0^{\frac{\pi}{4}} \sin\left(2x + \dfrac{\pi}{4}\right)dx$, expressing your answer in surd form.

. . . .

Answer

6 (a) (i) $\displaystyle\int \dfrac{1}{4x-7}\,dx = \dfrac{1}{4}\ln |4x-7| + c$

(ii) $\displaystyle\int e^{3x-1}\,dx = \dfrac{1}{3}e^{3x-1} + c$

(iii) $\displaystyle\int \dfrac{5}{(2x+3)^4}\,dx = 5\int (2x+3)^{-4}\,dx$

$\qquad\qquad\qquad\qquad = -\dfrac{5}{6}(2x+3)^{-3} + c$

(b) $\displaystyle\int_0^{\frac{\pi}{4}} \sin\left(2x + \dfrac{\pi}{4}\right)dx = \left[-\dfrac{1}{2}\cos\left(2x + \dfrac{\pi}{4}\right)\right]_0^{\frac{\pi}{4}}$

$\qquad\qquad\qquad = \left(-\dfrac{1}{2}\cos\dfrac{3\pi}{4}\right) - \left(-\dfrac{1}{2}\cos\dfrac{\pi}{4}\right)$

$$= \left(\frac{1}{2\sqrt{2}}\right) - \left(-\frac{1}{2\sqrt{2}}\right)$$

$$= \frac{2}{2\sqrt{2}}$$

$$= \frac{1}{\sqrt{2}}$$

$$= \frac{\sqrt{2}}{2}$$

$\cos\dfrac{\pi}{4} = \dfrac{1}{\sqrt{2}}$ and $\cos\dfrac{3\pi}{4} = -\dfrac{1}{\sqrt{2}}$

You must be able to write commonly used trigonometric values in surd form.

Multiply the top and bottom by √2 in order to remove the surd from the denominator.

7.3 Using definite integration to find the area between two curves

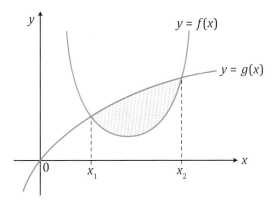

To find the area between two curves (e.g. the shaded area in the above diagram), first find the x-coordinates of the points of intersection by solving the two equations simultaneously.

Then integrate the curve $y = g(x)$ using the limits x_2 and x_1 and then integrate $y = f(x)$ between the same limits.

Finally subtract the area under $y = f(x)$ from the area under $y = g(x)$ to give a value for the shaded area.

Step by STEP

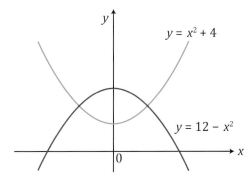

The diagram above shows a sketch of the curves $y = x^2 + 4$ and $y = 12 - x^2$. Find the area of the region bounded by the two curves.

Steps to take

1   Solve the equations of the two curves simultaneously so that the x-coordinates of the points of intersection can be found.

2   Integrate both equations between the limits of the x-coordinates of the points of intersection.

3   Subtract the area for the ∪-shaped curve from the ∩-shaped curve to give the required shaded area.

. . . .

Answer

$$y = x^2 + 4 \text{ and } y = 12 - x^2$$

The two equations are solved simultaneously to find the x-coordinates of the points of intersection.

Equating the y-values, we obtain

$$x^2 + 4 = 12 - x^2$$

$$2x^2 = 8$$

$$x = \pm 2$$

Hence the x coordinates of the points of intersection are 2 and −2.

Area under curve $y = 12 - x^2 = \displaystyle\int_{-2}^{2} \left(12 - x^2\right) dx$

$$= \left[12x - \frac{x^3}{3}\right]_{-2}^{2}$$

$$= \left[\left(12(2) - \frac{2^3}{3}\right) - \left(12(-2) - \frac{(-2)^3}{3}\right)\right]$$

$$= \frac{128}{3} \text{ square units}$$

Area under curve $y = x^2 + 4 = \displaystyle\int_{-2}^{2} \left(x^2 + 4\right) dx$

$$= \left[\frac{x^3}{3} + 4x\right]_{-2}^{2}$$

$$= \left[\left(\frac{2^3}{3} + 4(2)\right) - \left(\frac{(-2)^3}{3} + 4(-2)\right)\right]$$

$$= \frac{64}{3} \text{ square units}$$

Required area $= \dfrac{128}{3} - \dfrac{64}{3}$

$$= \frac{64}{3} \text{ square units}$$

BOOST

Grade ⇧⇧⇧⇧

Make sure you include the units of square units here.

7.4 Using integration as the limit of a sum

The graph below shows a section of the area of the curve $y = f(x)$ bounded by the x-axis and the two lines $x = a$ and $x = b$.

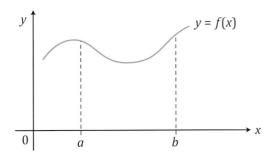

This area can be split into lots of small rectangular strips so the total area will be the sum of these strips. Each rectangular strip is drawn from the point P(x, y) as shown in the following diagram:

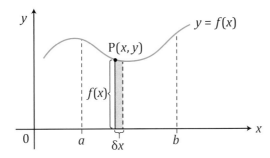

We assume that each strip is a rectangle having height $f(x)$ and width δx giving a small area δA given by:

$$\delta A \approx f(x)\delta x$$

Adding up the areas of all these thin strips from a to b we have

$$\text{Total area between } a \text{ and } b = \sum_{x=a}^{b} \delta A \approx \sum_{x=a}^{b} f(x)\delta x$$

As the width of each strip becomes very small, $\delta x \to 0$ and the total area now becomes:

$$\text{Total area} = \lim_{\delta x \to 0} \sum_{x=a}^{b} f(x)\delta x$$

As δx becomes smaller and smaller and the limit exists then we can define the definite integral as the limit of a sum and this can also be written as follows:

$$\int_{a}^{b} f(x)\delta x = \lim_{\delta x \to 0} \sum_{x=a}^{b} f(x)\delta x$$

We can regard integration as the process of adding up so it can be considered to be a summation. This is why we can use integration to find the area under a curve between two limits because it works out the area of a small portion and then sums this up over the region between the two limits.

7.5 Integration by substitution and integration by parts

Two new methods of integration are introduced here: integration by substitution and integration by parts.

Integration by substitution

Suppose you had to find $\int x(3x + 1)^5 \, dx$.

One method would be to expand the bracket and then multiply by x and simplify before finally integrating. There is quite a bit of work in this and there is a simpler method called integration by substitution. This method is shown here.

Suppose you are asked to integrate the following indefinite integral:

$$\int x(3x + 1)^5 \, dx$$

> Where there is a bracket raised to a power like this, you usually let u be equal to the contents of the bracket.

Let $u = 3x + 1$, so $\dfrac{du}{dx} = 3$. Hence $dx = \dfrac{du}{3}$

Notice that we can now replace $(3x + 1)^5$ by u^5, and dx by $\dfrac{du}{3}$.

There is still an x which needs to be replaced by an expression in terms of u and this can be obtained by rearranging $u = 3x + 1$ to give:

$$x = \frac{u - 1}{3}$$

> Note that the variable is now u rather than x.

The integral $\int x(3x + 1)^5 \, dx$ now becomes $\displaystyle\int \left(\frac{u - 1}{3}\right) u^5 \frac{du}{3}$

> Here the expression in u is integrated. As this is an indefinite integral, it is necessary to include the constant of integration.

Hence
$$\int \left(\frac{u - 1}{3}\right) u^5 \frac{du}{3} = \frac{1}{9} \int \left(u^6 - u^5\right) du$$

$$= \frac{1}{9}\left(\frac{u^7}{7} - \frac{u^6}{6}\right) + c$$

> Always remember to give the indefinite integral back in terms of x.

u can now be substituted back as $3x + 1$ to give:

$$\int x(3x + 1)^5 \, dx = \frac{1}{9}\left[\frac{(3x + 1)^7}{7} - \frac{(3x + 1)^6}{6}\right] + c$$

Suppose you have the same integral but this time it is a definite integral (i.e. an integral with limits), such as

$$\int_{-\frac{1}{3}}^{0} x(3x + 1)^5 \, dx$$

You would use the substitution to change the variable to u as before and change the limits so that they now apply to u and not x. This is done by substituting each limit for x into the following equation.

$u = 3x + 1$ so when $x = 0$, $u = 1$ and when $x = -\dfrac{1}{3}$, $u = 0$.

The integral now becomes

> There is no need to give the answer in terms of x: use the u limits.

$$\frac{1}{9}\int_{0}^{1}\left(u^6 - u^5\right) du = \frac{1}{9}\left[\frac{u^7}{7} - \frac{u^6}{6}\right]_{0}^{1} = -\frac{1}{378}$$

Integration by parts

Integration by parts is used when there is a product to integrate. The formula for integration by parts is obtained from the formula booklet and is as follows:

$$\int u \frac{dv}{dx} \, dx = uv - \int v \frac{du}{dx} \, dx$$

At times, integration by parts can be complicated. However in A2, one of the functions in the integrand is a polynomial, such as $x^2 + 2$, $4x + 3$. Hence in A2 we encounter integrals such as $\int x^2 \sin 2x \, dx$, $\int (3x + 2)\ln x \, dx$ and $\int x e^{3x} \, dx$.

The following rules are useful in A2.

Rule 1

When one of the functions is a polynomial and the other function is easily integrated, let u = polynomial,

$$\frac{dv}{dx} = \text{other function}$$

Rule 2

When one of the functions is a polynomial and the other function is not easily integrated,

Let $\qquad \dfrac{dv}{dx}$ = polynomial in x, and u = other function

Thus for $\int x e^{3x} \, dx$ we use Rule 1 since e^{3x} is easy to integrate, and we let $u = x$ and $\dfrac{dv}{dx} = e^{3x}$.

In contrast, for $\int x^2 \ln x$ we use Rule 2 since $\ln x$ is difficult to integrate, and we let $u = \ln x$ and $\dfrac{dv}{dx} = x^2$.

Examples

1 Find $\int x \cos 2x \, dx$.

. . . .

Answer

1 $\displaystyle \int u \frac{dv}{dx} \, dx = uv - \int v \frac{du}{dx} \, dx$

Let $u = x$ and $\dfrac{dv}{dx} = \cos 2x$

So $\dfrac{du}{dx} = 1$, $v = \dfrac{\sin 2x}{2}$

$\displaystyle \int x \cos 2x \, dx = \frac{1}{2} x \sin 2x - \int \frac{1}{2} \sin 2x \, dx$

$\displaystyle = \frac{x}{2} \sin 2x - \frac{1}{4} \cos 2x + c$

The formula for integration by parts can be obtained from the formula booklet.

Here $\cos 2x$ is easily integrated and we use Rule 1.

2   Find $\int x \ln x \, dx$.

- -

Answer

2    $\int u \dfrac{dv}{dx} \, dx = uv - \int v \dfrac{du}{dx} \, dx$

Here ln x is not easily integrated so we use Rule 2.

Let $u = \ln x$ and $\dfrac{dv}{dx} = x$

So that $\dfrac{du}{dx} = \dfrac{1}{x}$, $v = \dfrac{x^2}{2}$ Remember $\dfrac{d(\ln x)}{dx} = \dfrac{1}{x}$

Note the cancelling of x in the integral.

Hence $\int x \ln x \, dx = \ln x \left(\dfrac{x^2}{2}\right) - \int \dfrac{x^2}{2}\left(\dfrac{1}{x}\right) dx$

$= \dfrac{x^2}{2} \ln x - \int \dfrac{x}{2} \, dx$

$= \dfrac{x^2}{2} \ln x - \dfrac{x^2}{4} + c$

3   Find $\int \ln x \, dx$.

- -

Answer

3   In this case, there is only one function to be integrated and apparently integration by parts cannot be used. The trick is to regard ln x as ln $x \times 1$ and use Rule 2.

Then let $u = \ln x$

$\dfrac{dv}{dx} = 1$

Thus, $\dfrac{du}{dx} = \dfrac{1}{x}$, $v = x$

On substituting in

$\int u \dfrac{dv}{dx} \, dx = uv - \int v \dfrac{du}{dx} \, dx$

we obtain

$\int \ln x \, dx = (\ln x)x - \int x\left(\dfrac{1}{x}\right) dx$

$= x \ln x - \int 1 \, dx$

$= x \ln x - x + c$

Trigonometric substitution

In some cases we can make integration easier by using a trigonometric substitution.

The following example shows this technique.

Examples

1   Use the substitution $x = \sin \theta$ to show that

$\int_0^1 \sqrt{(1 - x^2)} \, dx = \dfrac{\pi}{4}$

Answer

Use the substitution given in the question.

1   Let $x = \sin \theta$ so $\dfrac{dx}{d\theta} = \cos \theta$ and $dx = \cos \theta \, d\theta$

When $x = 1$, $\sin \theta = 1$, so $\theta = \sin^{-1} 1 = \dfrac{\pi}{2}$

When $x = 0$, $\sin \theta = 0$, so $\theta = \sin^{-1} 0 = 0$

The limits are changed so that they apply to the new variable θ.

$$\int_0^1 \sqrt{(1 - x^2)} \, dx = \int_0^{\frac{\pi}{2}} \sqrt{(1 - \sin^2 \theta)} \cos \theta \, d\theta$$

Now $1 - \sin^2 \theta = \cos^2 \theta$

$$\int_0^{\frac{\pi}{2}} \sqrt{(1 - \sin^2 \theta)} \cos \theta \, d\theta = \int_0^{\frac{\pi}{2}} \sqrt{(\cos^2 \theta)} \cos \theta \, d\theta$$

$$= \int_0^{\frac{\pi}{2}} \cos \theta \cos \theta \, d\theta$$

$$= \int_0^{\frac{\pi}{2}} \cos^2 \theta \, d\theta$$

$$= \int_0^{\frac{\pi}{2}} \frac{1}{2} \left(1 + \cos 2\theta\right) d\theta$$

$$= \frac{1}{2} \int_0^{\frac{\pi}{2}} (1 + \cos 2\theta) \, d\theta$$

$$= \frac{1}{2} \left[\theta + \frac{1}{2} \sin 2\theta \right]_0^{\frac{\pi}{2}}$$

$$= \frac{1}{2} \left[\left(\frac{\pi}{2} + \frac{1}{2} \sin \pi \right) - \left(0 + \frac{1}{2} \sin 0 \right) \right]$$

$$= \frac{1}{2} \left[\left(\frac{\pi}{2} + 0 \right) - \left(0 + 0 \right) \right]$$

$$= \frac{\pi}{4}$$

This is a rearrangement of the double angle formula
$$\cos 2A = 2 \cos^2 A - 1$$
This formula is not in the formula booklet but you can derive it from other formulae which are in the formula booklet. Look back at Topic 4 if you are unsure as to how you would do this.

---

2   **(a)**   Find $\int (x + 3)e^{2x} \, dx$.

(b)   Use the substitution $u = 2 \cos x + 1$ to evaluate

$$\int_0^{\frac{\pi}{3}} \frac{\sin x}{\sqrt{(2 \cos x + 1)}} \, dx$$

Answer

This is the formula for integration by parts and is in the formula booklet. Use Rule 1 for the integration by parts.

2   **(a)** $\int u \dfrac{dv}{dx} \, dx = uv - \int v \dfrac{du}{dx} \, dx$

Let $u = (x + 3)$ and $\dfrac{dv}{dx} = e^{2x}$

$$\int (x + 3)e^{2x} \, dx = \left(x + 3 \right) \frac{1}{2} e^{2x} - \int \frac{1}{2} e^{2x}(1) \, dx$$

$$= \left(x + 3 \right) \frac{1}{2} e^{2x} - \frac{1}{4} e^{2x} + c$$

As $u = x + 3$, $\dfrac{du}{dx} = 1$

(b) Let $u = 2\cos x + 1$

$$\frac{du}{dx} = -2\sin x, \text{ so } dx = \frac{du}{-2\sin x}$$

When $x = \frac{\pi}{3}$, $u = 2\cos\frac{\pi}{3} + 1 = 2$

When $x = 0$, $u = 2\cos 0 + 1 = 3$

> Notice how $\sin x$ appears in both the numerator and the denominator and can therefore be cancelled.

$$\int_0^{\frac{\pi}{3}} \frac{\sin x}{\sqrt{(2\cos x + 1)}}\,dx = \int \frac{\sin x}{\sqrt{u}}\left(\frac{du}{-2\sin x}\right)$$

> Note the order of the limits. Do not interchange these limits.

$$= -\frac{1}{2}\int_3^2 u^{-\frac{1}{2}}\,du$$

$$= -\left[u^{\frac{1}{2}}\right]_3^2 = \left[-\sqrt{u}\right]_3^2 = \left[-\sqrt{2} - \left(-\sqrt{3}\right)\right] = -\sqrt{2} + \sqrt{3} = 0.318$$

(correct to 3 decimal places).

---

3 (a) Find $\int x^3 \ln x\,dx$.

(b) Use the substitution $u = 2x - 3$ to evaluate $\int_1^2 x(2x-3)^4\,dx$.

. . . .

Answer

> You must recognise that there is a product to be integrated so integration by parts is used. Obtain the formula from the formula booklet.

3 (a) $\int u\frac{dv}{dx}\,dx = uv - \int v\frac{du}{dx}\,dx$

Let $u = \ln x$ and $\frac{dv}{dx} = x^3$

> Notice that if you let $\frac{dv}{dx} = \ln x$ it would mean that $\ln x$ would need to be integrated. The integral of $\ln x$ is not known, so it is necessary to let $u = \ln x$ to avoid this (i.e. use Rule 2).

Then $\int \ln x\,(x^3)\,dx = \ln x\left(\frac{x^4}{4}\right) - \int\left(\frac{x^4}{4} \times \frac{1}{x}\right)dx$

$$= \frac{x^4}{4}\ln x - \int\frac{x^3}{4}dx$$

$$= \frac{x^4}{4}\ln x - \frac{x^4}{16} + c$$

(b) $u = 2x - 3$, $\frac{du}{dx} = 2$, so $dx = \frac{du}{2}$ and $x = \frac{u+3}{2}$

> Remember to change the limits so that they apply to the new variable u.

$$\int_1^2 x(2x-3)^4\,dx = \int\left(\frac{u+3}{2}\right)u^4\frac{du}{2} = \frac{1}{4}\int(u^5 + 3u^4)\,du$$

Change the limits using $u = 2x - 3$. When $x = 2$, $u = 1$ and when $x = 1$, $u = -1$.

Hence $\frac{1}{4}\int_{-1}^1 (u^5 + 3u^4)\,du = \frac{1}{4}\left[\frac{u^6}{6} + \frac{3u^5}{5}\right]_{-1}^1 = \frac{1}{4}\left[\left(\frac{1}{6} + \frac{3}{5}\right) - \left(\frac{1}{6} - \frac{3}{5}\right)\right] = \frac{3}{10}$

4 (a) Find $\int x \sin 2x \, dx$.

  (b) Use the substitution $u = 5 - x^2$ to evaluate

$$\int_0^2 \frac{x}{(5 - x^2)^3} \, dx.$$

· ·

Answer

4 (a) $\int x \sin 2x \, dx$

> This is a product and needs to be integrated by parts using the formula obtained from the formula booklet.
> Use Rule 1.

$$\int u \frac{dv}{dx} \, dx = uv - \int v \frac{du}{dx} \, dx$$

Let $u = x$ and $\dfrac{dv}{dx} = \sin^2 x$

$\dfrac{du}{dx} = 1$ and $v = \dfrac{-\cos 2x}{2}$

Then $\int x \sin 2x \, dx = x\left(-\dfrac{1}{2}\cos 2x\right) - \int\left(-\dfrac{1}{2}\cos 2x\right)(1)\,dx$

$$= -\frac{x}{2}\cos 2x + \frac{1}{4}\sin 2x + c$$

  (b) $u = 5 - x^2$, so $\dfrac{du}{dx} = -2x$, and $dx = \dfrac{du}{-2x}$

When $x = 2$, $u = 5 - 2^2 = 1$

$x = 0$, $u = 5 - 0 = 5$

> The limits are changed so that they apply now to u and not x.

Hence $\displaystyle\int_0^2 \frac{x}{(5 - x^2)^3} \, dx = \int_5^1 \frac{x}{u^3}\left(\frac{du}{-2x}\right)$

> Notice that the x in the numerator and denominator can be cancelled.

$$= -\frac{1}{2}\int_5^1 \frac{1}{u^3} \, du$$

$$= -\frac{1}{2}\int_5^1 u^{-3} \, du$$

$$= -\frac{1}{2}\left[\frac{u^{-2}}{-2}\right]_5^1$$

> Notice the order of the limits which must not be changed.

$$= -\frac{1}{2}\left[-\frac{1}{2u^2}\right]_5^1$$

$$= -\frac{1}{2}\left[\left(-\frac{1}{2}\right) - \left(-\frac{1}{50}\right)\right]$$

$$= -\frac{1}{2}\left(-\frac{1}{2} + \frac{1}{50}\right)$$

$$= \frac{6}{25}$$

Integration by substitution when you are not given the substitution to use

In the previous specification you were always given the substitution to use but now in some questions you will have to decide on the substitution to use. Deciding on the substitution is quite tricky and you need quite a bit of practice to become proficient. The following examples will guide you through the range of questions and types of substitution to use. Sometimes the substitution you use won't be the right one to use and you end up with something you can't integrate easily, so be prepared to abandon a particular substitution and use a different one.

Examples

1 Find, using a suitable substitution, $\int \dfrac{e^x}{1 + e^x} \, dx$

. . . .

Answer

Notice the differential of the bottom of the fraction is on the top so the integral is the ln of the bottom.

1 Let $u = e^x$ so $\dfrac{du}{dx} = e^x$ and $dx = \dfrac{du}{e^x} = \dfrac{du}{u}$

Hence using the substitution we obtain $\int \dfrac{e^x}{1 + e^x} \, dx = \int \dfrac{u}{1 + u} \times \dfrac{du}{u} = \int \dfrac{1}{1 + u} \, du$

$$= \ln(1 + u) + c = \ln(1 + e^x) + c$$

---

2 Find the value of $\displaystyle\int_0^{\frac{1}{2}} \dfrac{4x}{(1 + 2x)^4} \, dx$

. . . .

Answer

2 Let $u = 1 + 2x$, so $\dfrac{du}{dx} = 2$ and $dx = \dfrac{du}{2}$

Substituting these values into the integral, we obtain $\displaystyle\int \dfrac{4x}{u^4} \dfrac{du}{2}$

Notice that there is still an x in the integral; we need to find this in terms of u.

Using $u = 1 + 2x$ we have $x = \dfrac{u - 1}{2}$

Substituting this value into the integral, we obtain

$$\int 4 \dfrac{(u - 1)}{2u^4} \dfrac{du}{2} = \int \dfrac{(u - 1)}{u^4} \, du$$

Notice that we removed the limits. This is because the limits referred to x and not the new variable we have used, u. We now change the limits so they refer to u using the equation we used for the substitution.

When you have changed the variable, you must remember to change the limits.

As $u = 1 + 2x$, when $x = \dfrac{1}{2}$, $u = 2$ so the top limit becomes 2.

When $x = 0$, $u = 1$ so the bottom limit becomes 1.

Hence the integral now becomes

$$\int_1^2 \dfrac{(u - 1)}{u^4} \, du = \int_1^2 u^{-4}(u - 1) \, du$$

$$= \int_1^2 (u^{-3} - u^{-4})\, du$$

$$= \left[\frac{u^{-2}}{-2} - \frac{u^{-3}}{-3} \right]_1^2$$

$$= \left[-\frac{1}{2u^2} + \frac{1}{3u^3} \right]_1^2$$

$$= \left[\left(-\frac{1}{8} + \frac{1}{24} \right) - \left(-\frac{1}{2} + \frac{1}{3} \right) \right]$$

$$= \frac{1}{12}$$

BOOST

Grade ⇧⇧⇧⇧

Don't struggle working out fractions like this by finding common denominators – use a calculator instead.

3 Evaluate $\displaystyle\int_0^3 \frac{x}{\sqrt{x+1}}\, dx$

Answer

3   Let $u = x + 1$ so $\dfrac{du}{dx} = 1$ and $dx = du$

Hence, $\displaystyle\int \frac{x}{\sqrt{x+1}}\, dx = \int \frac{x}{u^{\frac{1}{2}}}\, du$

Now from $u = x + 1$, $x = u - 1$

The integral now becomes $\displaystyle\int \frac{(u-1)}{u^{\frac{1}{2}}}\, du = \int u^{-\frac{1}{2}}(u-1)\, du$

$$= \int \left(u^{\frac{1}{2}} - u^{-\frac{1}{2}} \right) du$$

Now when $x = 3$, $u = 3 + 1 = 4$ and when $x = 0$, $u = 1$

Hence

$$\int_1^4 \left(u^{\frac{1}{2}} - u^{-\frac{1}{2}} \right) du = \left[\frac{u^{\frac{3}{2}}}{\frac{3}{2}} - \frac{u^{\frac{1}{2}}}{\frac{1}{2}} \right]_1^4$$

$$= \left[\frac{2}{3} u^{\frac{3}{2}} - 2u^{\frac{1}{2}} \right]_1^4$$

$$= \left[\left(\frac{2}{3} \times 8 - 4 \right) - \left(\frac{2}{3} - 2 \right) \right]$$

$$= 2\frac{2}{3}$$

4 Evaluate $\displaystyle\int_0^{\frac{\pi}{2}} \frac{\cos x}{(4 + \sin x)^2}\, dx$

Answer

4   Let $u = 4 + \sin x$, so $\dfrac{du}{dx} = \cos x$ and $dx = \dfrac{du}{\cos x}$

Hence we have $\displaystyle\int \frac{\cos x}{u^2} \frac{du}{\cos x} = \int \frac{1}{u^2}\, du$

Remember if you use substitution, you need to change the limits so they apply to the new variable used.

When $x = \dfrac{\pi}{2}$, $u = 4 + \sin \dfrac{\pi}{2} = 5$

When $x = 0$, $u = 4 + \sin 0 = 4$

$$\int_4^5 \frac{1}{u^2}\, du = \int_4^5 u^{-2}\, du$$

$$= \left[-u^{-1}\right]_4^5$$

$$= \left[\left(-\frac{1}{5}\right) - \left(-\frac{1}{4}\right)\right]$$

$$= \frac{1}{20}$$

7.6 Integration using partial fractions

Converting a single algebraic fraction into two or more partial fractions was covered in Topic 2 of this book. In this section you will be turning an algebraic fraction into partial fractions and then integrating each of the partial fractions.

The following examples show the technique.

Examples

1   (a)   If $\dfrac{6 - 5x}{(1 - x)(2 - x)} \equiv \dfrac{A}{1 - x} + \dfrac{B}{2 - x}$, find the constants A and B.

(b)   Hence find $\displaystyle\int_{-1}^0 \dfrac{6 - 5x}{(1 - x)(2 - x)}\, dx$.

· ·

Answer

BOOST

Grade ⬆⬆⬆⬆

Always check the partial fractions are correct as a mistake may mean the partial fractions are harder to integrate.

1   (a) $\dfrac{6 - 5x}{(1 - x)(2 - x)} \equiv \dfrac{A}{1 - x} + \dfrac{B}{2 - x}$

$6 - 5x \equiv A(2 - x) + B(1 - x)$

Let $x = 2$ so $-4 = -B$, $B = 4$

Let $x = 1$,   so $A = 1$

Check the values of A and B are correct by substituting another value for x into the equation.

$x = 0$,   LHS $= \dfrac{6 - 5(0)}{(1)(2)} = 3$

RHS $= \dfrac{1}{1} + \dfrac{4}{2} = 3$

Hence LHS = RHS

(b) $\displaystyle\int_{-1}^0 \dfrac{6 - 5x}{(1 - x)(2 - x)}\, dx \equiv \int_{-1}^0 \left(\dfrac{1}{1 - x} + \dfrac{4}{2 - x}\right) dx$

When the numerator is the derivative of the denominator, the integral is ln of the denominator because

$\displaystyle\int \dfrac{f'(x)}{f(x)}\, du$ becomes $\displaystyle\int \dfrac{1}{u}\, du$

if we let $u = f(x)$.

$= \left[-\ln(1 - x) - 4\ln(2 - x)\right]_{-1}^0$

$= \left[(0 - 4\ln 2) - (-\ln 2 - 4\ln 3)\right]$

$= 4\ln 3 - 3\ln 2$

2 Given that $f(x) \equiv \dfrac{3x + 4}{(x + 3)(3x - 1)}$

(a) express $f(x)$ in terms of partial fractions,

(b) show that

$$\int_1^2 f(x)\,dx = \frac{1}{2}\ln\frac{25}{8}$$

Answer

2 (a) $\dfrac{3x + 4}{(x + 3)(3x - 1)} \equiv \dfrac{A}{x + 3} + \dfrac{B}{3x - 1}$

$$3x + 4 \equiv A(3x - 1) + B(x + 3)$$

Let $x = -3$ so $-5 = -10A$, hence $A = \dfrac{1}{2}$

Let $x = \dfrac{1}{3}$ so $5 = \dfrac{10}{3}B$, hence $B = \dfrac{3}{2}$

Partial fractions are $\dfrac{1}{2(x + 3)} + \dfrac{3}{2(3x - 1)}$

> Remember to check the partial fractions are correct by substituting in a different value of x and check that the left- and right-hand sides of the equation are equal.

(b) $\dfrac{1}{2}\displaystyle\int_1^2\left(\dfrac{1}{x + 3} + \dfrac{3}{3x - 1}\right)dx = \dfrac{1}{2}\Big[\ln(x + 3) + \ln(3x - 1)\Big]_1^2$

$$= \frac{1}{2}\Big[(\ln 5 + \ln 5) - (\ln 4 + \ln 2)\Big]$$

$$= \frac{1}{2}\ln\frac{25}{8}$$

> $\ln 5 + \ln 5 = \ln(5 \times 5) = \ln 25$
> and
> $\ln 4 + \ln 2 = \ln(4 \times 2) = \ln 8$
>
> $\ln 25 - \ln 8 = \ln\dfrac{25}{8}$

3 (a) Express $\dfrac{5x^2 + 6x + 7}{(x - 1)(x + 2)^2}$ as partial fractions.

(b) Using your answer from part (a), find

$$\int_2^3 \frac{5x^2 + 6x + 7}{(x - 1)(x + 2)^2}\,dx,\text{ giving your answer correct to two decimal places.}$$

Answer

3 (a) $\dfrac{5x^2 + 6x + 7}{(x - 1)(x + 2)^2} \equiv \dfrac{A}{(x - 1)} + \dfrac{B}{(x + 2)} + \dfrac{C}{(x + 2)^2}$

$$5x^2 + 6x + 7 \equiv A(x + 2)^2 + B(x - 1)(x + 2) + C(x - 1)$$

Let $x = 1$ so $18 = 9A$, giving $A = 2$

Let $x = -2$ so $15 = -3C$, giving $C = -5$

Let $x = 0$ so $7 = 8 - 2B + 5$, giving $B = 3$

$$\frac{5x^2 + 6x + 7}{(x - 1)(x + 2)^2} \equiv \frac{2}{(x - 1)} + \frac{3}{(x + 2)} - \frac{5}{(x + 2)^2}$$

> Note that there is a repeated linear factor of $(x + 2)^2$ here.

Check by letting $x = -1$

$$\text{LHS} = \frac{5x^2 + 6x + 7}{(x - 1)(x + 2)^2} = \frac{5(-1)^2 + 6(-1) + 7}{(-1 - 1)(-1 + 2)^2} = \frac{6}{-2} = -3$$

> This provides a useful check that the values of A, B and C are correct.

$$\text{RHS} = \frac{2}{(x - 1)} + \frac{3}{(x + 2)} - \frac{5}{(x + 2)^2} = \frac{2}{(-1 - 1)} + \frac{3}{(-1 + 2)} - \frac{5}{(-1 + 2)^2} = -1 + 3 - 5$$

$$= -3$$

LHS = RHS

Hence partial fractions are:

$$\frac{2}{(x-1)} + \frac{3}{(x+2)} - \frac{5}{(x+2)^2}$$

To integrate $-5(x+2)^{-2}$ increase the index by 1, divide by the derivative of the contents of the bracket and divide by the new index.

(b) $\displaystyle\int_2^3 \frac{5x^2 + 6x + 7}{(x-1)(x+2)^2}\, dx \equiv \int_2^3 \left(\frac{2}{(x-1)} + \frac{3}{(x+2)} - \frac{5}{(x+2)^2}\right) dx$

$$= \int_2^3 \left(\frac{2}{(x-1)} + \frac{3}{(x+2)} - 5(x+2)^{-2}\right) dx$$

$$= \left[2\ln(x-1) + 3\ln(x+2) + 5(x+2)^{-1}\right]_2^3$$

$$= \left[2\ln(x-1) + 3\ln(x+2) + \frac{5}{(x+2)}\right]_2^3$$

$$= \left[(2\ln 2 + 3\ln 5 + 1) - \left(2\ln 1 + 3\ln 4 + \frac{5}{4}\right)\right]$$

$$= 1.81 \text{ (correct to two decimal places)}$$

7.7 Analytical solution of first order differential equations with separable variables

First order differential equations are equations connecting x and y with $\frac{dy}{dx}$.

Here these equations can be solved to find an equation just connecting x and y by separating the variables and integrating. The following example shows this technique.

$$\int \frac{1}{g(y)}\, dy = \int f(x)\, dx$$

Examples

1 Solve the differential equation

$$\frac{dy}{dx} = 3xy^2$$

given that $x = 2$ when $y = 1$. Give the answer in the form $y = f(x)$

. . . .

Answer

1    $\dfrac{dy}{dx} = 3xy^2$

Separating the variables and integrating, we obtain

$$\int \frac{dy}{y^2} = \int 3x\, dx$$

You need to swap the equation around so that all the terms involving x are on one side of the equation and all the terms involving y are on the other side.

$$\int y^{-2}\, dy = \int 3x\, dx$$

$$-y^{-1} = \frac{3x^2}{2} + c$$

$$-\frac{1}{y} = \frac{3x^2}{2} + c$$

When $x = 2, y = 1$

Hence $-1 = 6 + c$, giving $c = -7$

then $\qquad -\dfrac{1}{y} = \dfrac{3x^2}{2} - 7$

> Multiplying both sides by $-2y$ to remove the fractions.

$$2 = -3x^2y + 14y$$

$$2 = y(14 - 3x^2)$$

> Note that the question asks for the equation to be given in the form $y = f(x)$

$$y = \dfrac{2}{14 - 3x^2}$$

---

2 (a) A cylindrical water tank has base area $4\,\text{m}^2$. The depth of the water at time t seconds is h metres. Water is poured in at the rate $0.004\,\text{m}^3$ per second.

Water leaks from a hole in the bottom at a rate of $0.0008h\,\text{m}^3$ per second.

Show that: $\qquad 5000\dfrac{dh}{dt} = 5 - h$

[Hint: the volume, V, of the cylindrical water tank is given by $V = 4h$.]

(b) Given that the tank is empty initially, find h in terms of t.

(c) Find the depth of the water in the tank when $t = 3600\,\text{s}$, giving your answer correct to 2 decimal places.

Answer

2 (a) The rate of change of volume = rate at which water comes in − rate at which the water leaves through the hole.

$$\dfrac{dV}{dt} = 0.004 - 0.0008h$$

> Look at the units to help you form the rate of change equation. Here we have the rate of change of volume in m^3 per second so this is a volume divided by a time.

Now as $V = 4h$, the rate of change of volume will be $\dfrac{dV}{dt}$ and the rate of change of height will be $\dfrac{dh}{dt}$.

Hence we can write $\dfrac{dV}{dt} = 4\dfrac{dh}{dt}$

So we have $\qquad\qquad\qquad 4\dfrac{dh}{dt} = 0.004 - 0.0008h$

Dividing through by 4 we obtain $\qquad \dfrac{dh}{dt} = 0.001 - 0.0002h$

Now multiply by 5000 to give $\quad 5000\dfrac{dh}{dt} \equiv 5 - h$

(b) Separating the variables and integrating, we obtain

$$5000\int \dfrac{dh}{5 - h} = \int dt$$

$$-5000 \ln (5 - h) = t + c \qquad\qquad (1)$$

> Note we need the derivative of the denominator (i.e. −1) on the top so we include a minus sign but we also need a minus sign in front of the ln to compensate.

Now $h = 0$ at $t = 0$ so

$$-5000 \ln 5 = c$$

Substituting c into equation (1) we have

$$-5000 \ln (5 - h) = t - 5000 \ln 5$$

Hence $\qquad\qquad\qquad \dfrac{t}{5000} = \ln\left(\dfrac{5}{5 - h}\right)$

This is done to remove the ln.

Taking exponentials of both sides

$$\frac{5}{5 - h} = e^{\frac{t}{5000}}$$

$$5 - h = 5e^{-\frac{t}{5000}}$$

$$h = 5 - 5e^{-\frac{t}{5000}}$$

(c) When $t = 3600$, $h = 5 - 5e^{-\frac{3600}{5000}}$

$$= 2.57 \text{ m (2 d.p.)}$$

Test yourself

1 Find

(a) $\displaystyle\int \frac{6}{5x + 1}\, dx$ [2]

(b) $\displaystyle\int \cos 7x\, dx$ [2]

(c) $\displaystyle\int \frac{4}{(3x + 1)^3}\, dx$ [2]

2 Find

(a) $\displaystyle\int \sin 4x\, dx$ [2]

(b) $\displaystyle\int \frac{1}{2x + 1}\, dx$ [2]

(c) $\displaystyle\int \frac{4}{(2x + 1)^5}\, dx$ [2]

3 Evaluate $\displaystyle\int_0^2 \frac{1}{(2x + 1)^3}\, dx$ [4]

4 Evaluate $\displaystyle\int_0^3 \frac{1}{5x + 2}\, dx$, expressing your answer as a logarithm. [4]

5 Find $\displaystyle\int (2x + 1)e^{2x}\, dx$ [4]

6 By using the substitution $x = 2 \sin \theta$, evaluate

$\displaystyle\int_0^1 \sqrt{(4 - x^2)}\, dx$ giving your answer correct to three decimal places. [5]

7 Solve the equation
$$\frac{dy}{dx} = \frac{y}{x + 2}$$
given that $y = 2$ when $x = 0$ [5]

Summary

Check you know the following facts:

Integration of $x^n (n \neq -1)$, e^{kx}, $\frac{1}{x}$, $\sin kx$, $\cos kx$

$$\int x^n \, dx = \frac{x^{n+1}}{n+1} + c$$

$$\int e^{kx} \, dx = \frac{e^{kx}}{k} + c$$

$$\int \frac{1}{x} \, dx = \ln|x| + c$$

$$\int \sin kx \, dx = -\frac{1}{k}\cos kx + c$$

$$\int \cos kx \, dx = \frac{1}{k}\sin kx + c$$

You will be required to remember these results as they are not given in the formula booklet.

Integration of $(ax + b)^n (n \neq -1)$, e^{ax+b}, $\frac{1}{ax+b}$, $\sin(ax+b)$, $\cos(ax+b)$

$$\int (ax+b)^n \, dx = \frac{(ax+b)^{n+1}}{(n+1)a} + c \qquad (n \neq -1)$$

$$\int e^{ax+b} \, dx = \frac{e^{ax+b}}{a} + c$$

$$\int \frac{1}{ax+b} \, dx = \frac{1}{a}\ln|ax+b| + c$$

$$\int \sin(ax+b) \, dx = \frac{-\cos(ax+b)}{a} + c$$

$$\int \cos(ax+b) \, dx = \frac{\sin(ax+b)}{a} + c$$

You will be required to remember these results as they are not given in the formula booklet.

Integration by substitution and integration by parts

An integral of the type $\int f(x) \, dx$ is converted into the integral $\int f(x)\frac{dx}{du} \, du$, where x is replaced by a given substitution. In the case of definite integrals, the x limits are converted into u limits by means of the given substitution.

Integration by parts

Integration by parts is used when there is a product to integrate.

$$\int u \frac{dv}{dx} \, dx = uv - \int v \frac{du}{dx} \, dx$$

Integration using partial fractions

Single fractions such as $\frac{5x+3}{(x+3)(x-1)}$ can be converted into partial fractions $\left(\frac{3}{(x+3)}+\frac{2}{(x-1)}\right.$ in this case$\left.\right)$ so that each of the resulting partial fractions can be integrated.

In many cases the answer to these questions involves the use of ln.

Analytical solution of first order differential equations with separable variables

Equations of the type

$$\frac{dy}{dx}=f(x)\,g(y)$$

can be solved by separating the variables and integrating both sides of the resulting equation, i.e.

$$\int \frac{1}{g(y)}\,dy = \int f(x)\,dx$$

8 Numerical methods

Introduction

Numerical methods can be used to find the roots of equations/functions which can be put into the form $f(x) = 0$. These methods can be used to find the roots of equations where it would be difficult to find them using the algebraic techniques you have already covered. Numerical methods can also provide approximate solutions to equations using iterative methods. Numerical methods can also be used to find the approximate area under a curve.

This topic covers the following:

8.1 Location of roots of $f(x) = 0$, considering changes of sign of $f(x)$

8.2 Sequences generated by a simple recurrence relation of the form $x_{n+1} = f(x_n)$

8.3 Solving equations approximately using simple iterative methods

8.4 Solving equations using the Newton–Raphson method and other recurrence relations of the form $x_{n+1} = g(x_n)$

8.5 Numerical integration of functions using the Trapezium rule

8.1 Location of roots of $f(x) = 0$ considering changes of sign of $f(x)$

If $f(x)$ can take any value between a and b, then if there is a change of sign between $f(a)$ and $f(b)$, then a root of $f(x) = 0$ lies between a and b.

It is assumed that the graph of $y = f(x)$ is an unbroken line between $x = a$, $x = b$.

Examples

1   Show that the equation

$$9x^3 - 9x + 1 = 0$$

has a root α between 0 and 0.2 .

. . . .

Answer

1   Let $f(x) = 9x^3 - 9x + 1$

$$f(0) = 1$$

$$f(0.2) = 9(0.2)^3 - 9(0.2) + 1$$

$$= 0.072 - 1.8 + 1 = -0.728$$

$$f(0.2) = -0.728$$

As there is a sign change, the root α must lie between 0 and 0.2 .

> Insert both values in turn into the function and if there is a sign change, then the root lies between the two values.

---

2   Show that the function given by

$$f(x) = x^4 + 4x^2 - 32x + 5$$

has a stationary point when x satisfies the equation

$$x^3 + 2x - 8 = 0$$

Show that the equation

$$x^3 + 2x - 8 = 0$$

has a root α between 1 and 2.

The recurrence relation

$$x_{n+1} = (8 - 2x_n)^{\frac{1}{3}}$$

with $x_0 = 1.7$ may be used to find α. Find and record the values of x_1, x_2, x_3, x_4. Write down the value of x_4 correct to four decimal places.

. . . .

Answer

2   The stationary point occurs when $f' = 0$, i.e. when

$$4x^3 + 8x - 32 = 0$$

$$\therefore \quad x^3 + 2x - 8 = 0$$

Let $f(x) = x^3 + 2x - 8$

$$f(1) = (1)^3 + 2(1) - 8 = -5$$

$$f(2) = (2)^3 + 2(2) - 8 = 4$$

The sign change indicates a root α between 1 and 2.

> The two values between which the root is supposed to lie are entered in turn into the function.
>
> A sign change indicates that the root lies between the two values.

$x_0 = 1.7$

$$x_1 = (8 - 2x_0)^{\frac{1}{3}} = (8 - 2\,(1.7))^{\frac{1}{3}} = 1.663103499$$

$$x_2 = (8 - 2x_1)^{\frac{1}{3}} = (8 - 2\,(1.663103499))^{\frac{1}{3}} = 1.671949509$$

$$x_3 = (8 - 2x_2)^{\frac{1}{3}} = (8 - 2\,(1.671949509))^{\frac{1}{3}} = 1.669837194$$

$$x_4 = (8 - 2x_3)^{\frac{1}{3}} = (8 - 2\,(1.669837194))^{\frac{1}{3}} = 1.670342073$$

Hence $x_4 = 1.6703$ (correct to 4 decimal places).

> Always round your final answer to the required number of decimal places or significant figures.

How change of sign methods can fail

The graph of $y = f(x)$ shown below cuts the x-axis between 1 and 2 so it has a root between 1 and 2. The graph is a continuous curve and the diagram shows part of this curve between $x = 1$ and $x = 2$. A change of sign method can be used because the graph is continuous.

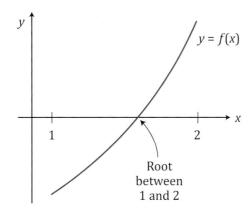

The following graph showing $f(x) = \frac{1}{x}$ is not continuous for all values of x.

The value of x cannot be zero as the curve approaches, but never touches, the y-axis as the y-axis is an asymptote.

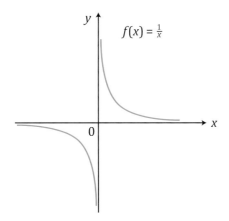

If you looked at a point either side of the y-axis (e.g. $x = -1$ and 1) then $f(-1) = -1$ and $f(1) = 1$ so there is a change of sign, falsely indicating that there is a root between $x = -1$ and 1. However, we can see by looking at the graph that the curve does not cut the x-axis, so no such root exists.

Here is another way a change of sign method would fail. Look at the following graph of a cubic function.

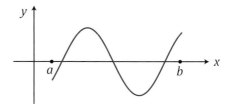

Suppose we looked at the sign of the function at the points $x = a$ and $x = b$.

$f(a)$ is negative and $f(b)$ is positive. There is a sign change indicating there is a root between a and b. However, there are in fact three roots in this interval. It would be necessary to investigate the signs for the function at intermediate values between a and b to fully locate all the roots.

Example

1   (a)   Show that the function given by

$$f(x) = x^3 + 3x^2 - 6x + 1$$

has a stationary point when x satisfies the equation:

$$x^2 + 2x - 2 = 0$$

(b)   Show that the equation $x^2 + 2x - 2 = 0$ has a root between 0 and 1.

. . . .

Answer

1   (a)   The stationary point occurs when $f' = 0$, i.e. when

$$3x^2 + 6x - 6 = 0$$

Therefore $x^2 + 2x - 2 = 0$

The two values between which the root is supposed to lie are entered in turn into the function.

A sign change indicates that the root lies between the two values.

(b)   Let $f(x) = x^2 + 2x - 2$

$$f(1) = (1)^2 + 2(1) - 2 = 1$$

$$f(0) = (0)^2 + 2(0) - 2 = -2$$

The sign change indicates a root between 0 and 1.

8.2 Sequences generated by a simple recurrence relation of the form $x_{n+1} = f(x_n)$

The following relation is called a recurrence relation,

$$x_{n+1} = x_n^3 + \frac{1}{9}$$

This recurrence relation can be used to generate a sequence by substituting a starting value called x_0 into the relation to calculate the next term in the sequence, called x_1. The value of x_1 is then substituted for x_n into the relation to calculate x_2. The process is repeated until the desired number of terms of the sequence have been found.

Examples

1 A sequence is generated using the recurrence relation

$$x_{n+1} = x_n^3 + \frac{1}{9}$$

Starting with $x_0 = 0.1$, find and record x_1, x_2, x_3.

. . . .

Answer

1 $x_0 = 0.1$

$$x_1 = x_0^3 + \frac{1}{9} = (0.1)^3 + \frac{1}{9} = 0.1121111111$$

$$x_2 = x_1^3 + \frac{1}{9} = (0.1121111111)^3 + \frac{1}{9} = 0.1125202246$$

$$x_3 = x_2^3 + \frac{1}{9} = (0.1125202246)^3 + \frac{1}{9} = 0.1125357073$$

> Do not round off any of your values. Write down the full calculator display.

---

2 Show that the equation

$$4x^3 - 2x - 5 = 0$$

has a root α between 1 and 2.

The recurrence relation

$$x_{n+1} = \left(\frac{2x_n + 5}{4}\right)^{\frac{1}{3}}$$

with $x_0 = 1.2$, may be used to find α. Find and record the values of x_1, x_2, x_3, x_4. Write down the value of x_4 correct to five decimal places and prove that this value is the value of α correct to five decimal places.

. . . .

Answer

2 Let $f(x) = 4x^3 - 2x - 5$

$$f(1) = 4(1)^3 - 2(1) - 5 = -3$$

$$f(2) = 4(2)^3 - 2(2) - 5 = 23$$

There is a change in sign between 1 and 2. Hence there is a root between these two values.

$$x_{n+1} = \left(\frac{2x_n + 5}{4}\right)^{\frac{1}{3}}$$

$x_0 = 1.2$

$$x_1 = \left(\frac{2x_0 + 5}{4}\right)^{\frac{1}{3}} = \left(\frac{2(1.2) + 5}{4}\right)^{\frac{1}{3}} = 1.227601026$$

$$x_2 = \left(\frac{2x_1 + 5}{4}\right)^{\frac{1}{3}} = \left(\frac{2(1.227601026) + 5}{4}\right)^{\frac{1}{3}} = 1.230645994$$

$$x_3 = \left(\frac{2x_2 + 5}{4}\right)^{\frac{1}{3}} = \left(\frac{2(1.230645994) + 5}{4}\right)^{\frac{1}{3}} = 1.230980996$$

> Both values between which the solution lies are entered in turn for x. If there is a sign change, then the solution lies between these two values.

> Do not round off these numbers yet.

$$x_4 = \left(\frac{2x_3 + 5}{4}\right)^{\frac{1}{3}} = \left(\frac{2(1.230980996) + 5}{4}\right)^{\frac{1}{3}} = 1.231017841$$

$x_4 = 1.23102$ (correct to five decimal places)

Here we need to look at the value of α either side of the value of $x_4 = 1.23102$ (i.e. at 1.231015 and at 1.231025).

$$f(1.231015) = 4(1.231015)^3 - 2(1.231015) - 5 = -0.000119667$$

$$f(1.231025) = 4(1.231025)^3 - 2(1.231025) - 5 = 0.000042182$$

As there is a sign change, $\alpha = 1.23102$ correct to five decimal places.

---

3 **(a)** Sketch the graphs of $y = x^3$ and $y = x + 2$. Deduce the number of real roots of the equation

$$x^3 - x - 2 = 0$$

(b) The cubic equation $x^3 - x - 2 = 0$ has a root α between 1 and 2.

The recurrence relation

$$x_{n+1} = (x_n + 2)^{\frac{1}{3}}$$

with $x_0 = 1.5$, can be used to find α.

Calculate x_4, giving your answer correct to three decimal places.
Prove that this value is also the value of α correct to three decimal places.

. . . .

Answer

3 **(a)**

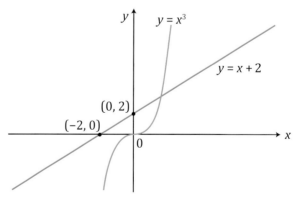

There is a real root of $x^3 - x - 2 = 0$ where the graphs of $y = x^3$ and $y = x + 2$ intersect. The graphs intersect once, so there is one real root of the equation $x^3 - x - 2 = 0$.

(b) $x_1 = 1.5182945$

$x_2 = 1.5209353$

$x_3 = 1.5213157$

$x_4 = 1.5213705 \approx 1.521$ (correct to three decimal places)

Check value of $x^3 - x - 2$ for $x = 1.5205, 1.5215$

x	$f(x)$
1.5205	−0.0052
1.5215	0.0007

Since there is a change of sign, the root is 1.521 correct to three decimal places.

Staircase diagrams

If you have an equation $g(x) = 0$ that can be written in the form $x = g(x)$, then the recurrence relation $x_{n+1} = g(x_n)$ for a chosen starting value (called x_0) may produce a convergent sequence which will lead to one of the roots of the equation $g(x) = 0$.

You can see how this works by looking at the following graph:

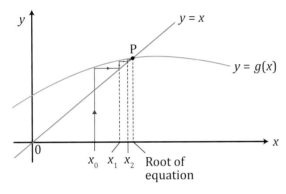

The curve $y = g(x)$ intersects the line $y = x$ at the point P. The x-coordinate of P is the root of the equation $x = g(x)$.

This can be solved using the recurrence relation $x_{n+1} = g(x_n)$.

A starting value x_0 is chosen and where the vertical line for this value cuts the curve we draw a horizontal line to the line $y = x$. The x-coordinate of this point gives the value of x_1. A vertical line is drawn to touch the curve and at this point a horizontal line is drawn. The x-coordinate of where the horizontal line meets the line $y = x$, is the value of x_2. The more iterations (or steps) we perform, the nearer the value of x gets to the value for the root.

In this case the sequence produced by the iteration converges to a steady value but this does not always happen as the following graph shows.

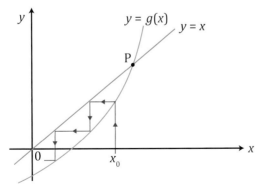

In the above graph you can see that the staircase diagram shows the values of x_1, x_2, etc., are moving further away from the x-value of the point of intersection at P. This is because the sequence diverges. Hence we cannot use the staircase diagram to find a root of a diverging sequence.

Cobweb diagrams

A cobweb diagram allows you to see whether the iteration is convergent or not. A convergent iteration starts at x_0 and the subsequent iterations x_1, x_2, x_3, ... converge to a value that is closer and closer to the root (i.e. the point of intersection of the two graphs).

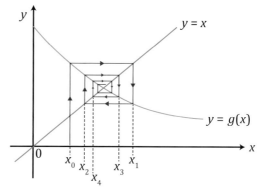

To draw a cobweb diagram to see how an iteration behaves, follow these steps:

1   Using the same set of axes, draw the graphs of $y = x$ and $y = g(x)$.

2   Start at a point on the x-axis x_0.

3   Draw a line vertically up to meet the graph of $y = g(x)$.

4   Draw a line horizontally to meet the graph of $y = x$. As this is the first iteration, the x-value is x_1, and when this step is repeated you will obtain x_2, x_3, x_4, ...

5   Return to step 3. The values of x_1, x_2, x_3, ... should get closer to the root of the equation (i.e. where the graphs intersect) if the iteration converges (i.e. gets nearer and nearer to a steady value). However if the iteration gets further and further away from a fixed value the iteration is diverging and the root cannot be found using this method.

Example

1   (a)   On the same axes, sketch the graphs of $y = x$ and $y = \cos x$ for $0 \leq x \leq \frac{\pi}{2}$.

   (b)   (i)   Hence, show that the equation $\cos x = x$ has only one root.

      (ii)   Estimate the root correct to one decimal place.

      (iii)   Using the value for your root as x_0, use the recurrence relation $x_{n+1} = \cos x_n$ to find the root correct to two decimal places.

Answer

1   (a)

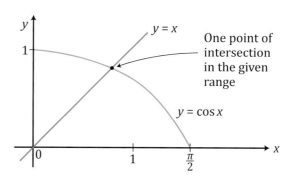

(b) (i) The curve and the line intersect at only one point in the given range.

    (ii) $x = 0.7$ radians (this can be read off from the graph)

    (iii) $x_{n+1} = \cos x_n$ and $x_0 = 0.7$

$$x_1 = \cos 0.7 = 0.76484\ldots$$

$$x_2 = \cos 0.76484\ldots = 0.72149\ldots$$

$$x_3 = \cos 0.72149\ldots = 0.75082\ldots$$

$$x_4 = \cos 0.75082\ldots = 0.73112\ldots$$

$$x_5 = \cos 0.73112\ldots = 0.74442\ldots$$

$$x_6 = \cos 0.74442\ldots = 0.73548\ldots$$

$$x_7 = \cos 0.73548\ldots = 0.74151\ldots$$

The last three answers give a constant value of 0.74 correct to two decimal places.

Hence the root = 0.74 radians correct to two decimal places

> To check the root is correct, change your calculator to radians, and then find the value of cos 0.7. It should be near to 0.7.

8.3 Solving equations approximately using simple iterative methods

To find an approximate solution to an equation by iteration, you first start with an approximate value to the root and then substitute this into a recurrence relation. The result is then substituted back into the recurrence relation and the process is repeated until a value to the desired degree of accuracy is obtained.

This technique can be seen in the following example.

Examples

1 You may assume that the equation $6x^4 + 7x - 3 = 0$ has a root α between 0 and 1.

The recurrence relation with $x_0 = 0.4$ can be used to find the root.

$$x_{n+1} = \frac{3 - 6x_n^4}{7}$$

Find and record the values of x_1, x_2, x_3, x_4. Write down the value of x_4 correct to four decimal places and show this is the value of α correct to four decimal places.

* *

Answer

1 $x_{n+1} = \dfrac{3 - 6x_n^4}{7}$

> This is the iterative formula and is given in the question. This is the formula into which the starting value and subsequent values are entered.

$x_0 = 0.4$

$$x_1 = \frac{3 - 6x_0^4}{7} = \frac{3 - 6(0.4)^4}{7} = 0.406628571$$

> x_0 is the starting value and is given in the question. This value is substituted into the formula to obtain the next value x_1

$$x_2 = \frac{3 - 6x_1^4}{7} = \frac{3 - 6(0.406628571)^4}{7} = 0.405137517$$

$$x_3 = \frac{3 - 6x_2^4}{7} = \frac{3 - 6(0.405137517)^4}{7} = 0.405479348$$

$$x_4 = \frac{3 - 6x_3^4}{7} = \frac{3 - 6(0.405479348)^4}{7} = 0.405401314$$

Hence $x_4 = 0.4054$ (correct to four decimal places).

Let $f(x) = 6x^4 + 7x - 3$

$$f(0.40535) = 6(0.40535)^4 + 7(0.40535) - 3 = -5.66 \times 10^{-4}$$

$$f(0.40545) = 6(0.40545)^4 + 7(0.40545) - 3 = 2.94 \times 10^{-4}$$

As there is a sign change, $\alpha = 0.4054$ correct to four decimal places.

Here we need to look at the value of α either side of the value of $x_4 = 0.4054$ (i.e. at 0.40535 and 0.40545).

If there is a sign change when these values are put into $6x^4 + 7x - 3$, then α is the root correct to 4 d.p.

---

2   Show that the equation

$$x - \sin x - 0.2 = 0,$$

where x is measured in radians, has a root α between 1 and 2.

The recurrence relation

$$x_{n+1} = \sin x_n + 0.2$$

with $x_0 = 1.1$ can be used to find α. Find and record the values of x_1, x_2, x_3, x_4.

Write down the value of x_4 correct to three decimal places and show this is the value of α correct to three decimal places.

· ·

Answer

2 | x | $x - \sin x - 0.2$ |
|---|---|
| 1 | -4.15×10^{-2} |
| 2 | 8.9×10^{-1} |

Since there is a change of sign, there is a root between 1 and 2.

$$x_1 = \sin(1.1) + 0.2 = 1.09120736$$

$$x_2 = \sin(1.09120736) + 0.2 = 1.087184654$$

$$x_3 = \sin(1.087184654) + 0.2 = 1.085321346$$

$$x_4 = 1.084453409$$

Hence $x_4 \approx 1.084$ (correct to three decimal places)

Now we check 1.0835 and 1.0845 in the original equation (not in the recurrence relation).

x	$x - \sin x - 0.2$
1.0835	-1.02×10^{-4}
1.0845	4.3×10^{-4}

Since there is a change of sign, the root is 1.084 correct to three decimal places.

8.4 Solving equations using the Newton–Raphson method and other recurrence relations of the form $x_{n+1} = g(x_n)$

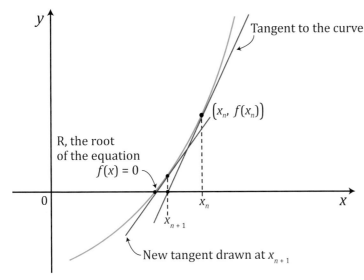

The Newton–Raphson method is used to solve equations of the form $f(x) = 0$.

The above graph shows part of the curve $y = f(x)$ and the x-coordinate of point R is the root of the equation $f(x) = 0$ (i.e. it is the x-coordinate of where the curve cuts the x-axis).

We start with a point having x-coordinate x_n and draw a tangent at this point. This tangent cuts the x-axis at x_{n+1}. At the point on the curve where the x-coordinate is x_{n+1} another tangent is drawn and this cuts the x-axis at x_{n+2}. The more times this process is repeated, the nearer the x-value gets to the correct value for the root of the equation (i.e. the x-coordinate of R).

The iteration equation we use to solve $f(x) = 0$ is:

$$x_{n+1} = x_n - \frac{f(x_n)}{f'(x_n)}$$

Note that $f'(x_n)$ is the derivative of the original function.

Step by STEP

The equation $1 + 5x - x^4 = 0$ has a positive root α.

(a) Show that α lies between 1 and 2.

(b) Use the iterative sequence based on the arrangement,

$$x = \sqrt[4]{1 + 5x}$$

with starting value 1.5 to find α correct to two decimal places.

(c) Use the Newton–Raphson method to find α correct to six decimal places.

Steps to take

1   Let $f(x) = 1 + 5x - x^4$ and substitute each value in turn. Zero should lie between the two values obtained.

2 Write the iterative sequence using x_n and x_{n+1}.

Use $x_0 = 1.5$ and start the iteration, finishing when the first two decimal places do not change.

3 Write the equation as $f(x) = \dots$. Differentiate the function to find $f'(x)$.

Obtain the formula from the formula booklet.

Keep substituting values in until the first six numbers after the decimal place remain constant.

Answer

(a) $f(x) = 1 + 5x - x^4$

$f(1) = 1 + 5(1) - (1)^4 = 5$

$f(2) = 1 + 5(2) - (2)^4 = -5$

As there is a change in sign, a root must exist between 1 and 2.

(b) $$x_{n+1} = \sqrt[4]{1 + 5x_n}$$

$$x_1 = \sqrt[4]{1 + 5x_0}$$

$x_0 = 1.5$ so $x_1 = \sqrt[4]{1 + 5(1.5)} = 1.707476485$

$x_2 = 1.757346089$

$x_3 = 1.768721303$

$x_4 = 1.771285475$

$x_5 = 1.771861948$

Hence root $\alpha \approx 1.77$

Make sure you do not round your answer. You should write down all the numbers in your calculator display.

(c) $$x_{n+1} = x_n - \frac{f(x_n)}{f'(x_n)}$$

This formula is obtained from the formula booklet.

$$x_{n+1} = x_n - \frac{1 + 5x_n - x_n^4}{5 - 4x_n^3}$$

$x_0 = 1.5$

$$x_1 = x_0 - \frac{1 + 5x_0 - x_0^4}{5 - 4x_0^3} = 1.5 - \frac{1 + 5(1.5) - (1.5)^4}{5 - 4(1.5)^3} = 1.904411765$$

$x_2 = 1.788115338$

$x_3 = 1.772305155$

$x_4 = 1.772029085$

$x_5 = 1.772029002$

Root $\alpha \approx 1.772029$

The question requires you to find the root α correct to six decimal places. You can see that x_4 and x_5 give the same value to six decimal places so we stop the iteration and use this value as the root.

How the Newton–Raphson method can fail

The Newton–Raphson method uses the gradient of the line as a starting point for the iteration but it does not work for every function. The Newton–Raphson method can fail if the gradient of the function is too small.

Look at the following example where the Newton–Raphson method fails to work.

Here we will find a non-zero solution to $f(x) = x - 2 \sin x$.

$$f'(x) = 1 - 2 \cos x$$

$$x_{n+1} = x_n - \frac{f(x_n)}{f'(x_n)} = x_n - \frac{x_n - 2 \sin x_n}{1 - 2 \cos x_n}$$

Suppose we start with $x_0 = 1.1$

The values of x_1 to x_6 are as follows:

$x_1 = 8.453$

$x_2 = 5.256$

$x_3 = 203.384$

$x_4 = 118.019$

$x_5 = -87.471$

$x_6 = -203.637$

You can see clearly the values are all over the place and we will have to do a very large number of iterations before you find the root.

> All of these numbers are given to 3 decimal places.

> **BOOST**
> **Grade** ⇧⇧⇧⇧
>
> You may be asked in questions about when numerical methods fail. The Newton–Raphson method can fail if the gradient of the function is too small.

8.5 Numerical integration of functions using the Trapezium rule

You already know that when you find the integral of a function $f(x)$ between two limits a and b, you are finding the area under the curve between those values. It therefore follows that if you have a known function $f(x)$ and you have a value for the area between a and b, then this value will be equal to the integral between those two limits.

Hence, if we know the function and can't find its area using a different method, then we can find the integral of the function.

Approximation of the area under a curve using the Trapezium rule

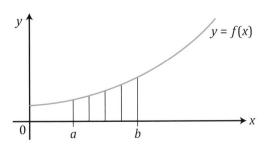

> Each y-value is called an ordinate. The ordinates correspond to x-values that divide the area into vertical strips of equal width. The number of strips is always one less than the number of ordinates used.

Look at the curve above. Suppose we want to find the area under the curve between $x = a$ and $x = b$. The area under the curve between these two points, is divided into strips (4 in this case) of equal width. Each strip can be approximated to a trapezium by making the assumption that there is a straight line at the top of the strip rather than a curve. By working out the areas of all the trapezia (the plural for trapezium) and then adding them together we get the approximate area under the curve. The greater the number of strips used, the more accurate the approximation to the true area becomes.

The approximate area under a curve may be found using a formula called the Trapezium rule. The Trapezium rule is:

$$\int_{a}^{b} y \, dx \approx \frac{1}{2} h \{ (y_0 + y_n) + 2(y_1 + y_2 + \ldots + y_{n-1}) \}, \qquad \text{where } h = \frac{b-a}{n}$$

h is the width of the strips used to estimate the area.

n is the number of strips used. n is always one less than the number of ordinates used. For example, if 5 ordinates are being used to estimate the area, then the number of strips used, n, will be 4.

y_0 and y_n are the first and last ordinates respectively.

$y_1, y_2, \ldots, y_{n-1}$, are the other ordinates between the first and the last.

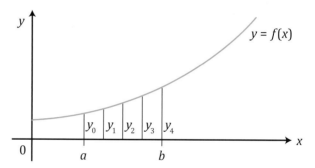

Overestimating and underestimating areas using the Trapezium rule

The Trapezium rule will overestimate the area if the tops of the trapezia are above the curve and underestimate the area if they are below the curve.

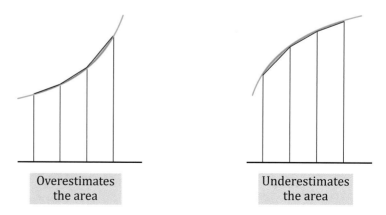

Increasing the accuracy of the estimate of the area using the Trapezium rule

The accuracy of the estimate of the area obtained using the Trapezium rule can be increased by increasing the number of ordinates or strips used.

The use of the Trapezium rule is explained using the following examples.

Examples

1 Use the Trapezium rule with five ordinates to find an approximate value for the integral

$$\int_0^1 \sqrt{\frac{1}{1 + x^2}}\,dx$$

Show your working and give your answer correct to three decimal places.

Answer

1 $h = \dfrac{b - a}{n} = \dfrac{1 - 0}{4} = 0.25$

$$\int_0^1 \sqrt{\frac{1}{1 + x^2}}\,dx \approx \frac{1}{2}h\{(y_0 + y_4) + 2(y_1 + y_2 + y_3)\}$$

When $x = 0$, $y_0 = \sqrt{\dfrac{1}{1 + (0)^2}} = 1$

        $x = 0.25$, $y_1 = \sqrt{\dfrac{1}{1 + (0.25)^2}} = 0.970143$

        $x = 0.5$, $y_2 = \sqrt{\dfrac{1}{1 + (0.5)^2}} = 0.894427$

        $x = 0.75$, $y_3 = \sqrt{\dfrac{1}{1 + (0.75)^2}} = 0.8$

        $x = 1$, $y_4 = \sqrt{\dfrac{1}{1 + (1)^2}} = 0.707107$

$$\int_0^1 \sqrt{\frac{1}{1 + x^2}}\,dx \approx \frac{1}{2} \times 0.25\{(1 + 0.707107) + 2(0.970143 + 0.894427 + 0.8)\}$$

$$\approx 0.87953$$

$$\approx 0.880 \text{ (to 3 decimal places)}.$$

> *h gives the width of the strips. b and a are the top and bottom limits of the integral. n, the number of strips used, is one less than the number of ordinates.*

> *This formula and also the formula for h are obtained from the formula booklet.*

> Starting from the lower limit (i.e. the value of *a*) and working in steps of *h* (i.e. 0.25 here) the values of *x* are substituted into the expression inside the integral. This gives the ordinates y_0, y_1, etc., which can then be entered into the formula for the Trapezium rule.

> Notice the question asks to give the answer correct to three decimal places. You must give the intermediate working to more decimal places and then give the final answer to three decimal places.

2 Use the Trapezium rule with five ordinates to find an approximate value for the integral

$$\int_1^2 \sqrt{1 + \frac{1}{x}}\,dx$$

Show your working and give your answer correct to three decimal places.

An answer on its own with no working will earn no marks.

Answer

2   $h = \dfrac{b-a}{n} = \dfrac{2-1}{4} = 0.25$

$$\int_1^2 \sqrt{1 + \frac{1}{x}}\, dx \approx \frac{1}{2} h\left\{(y_0 + y_n) + 2(y_1 + y_2 + \ldots + y_{n-1})\right\}$$

When $x = 1$, $y_0 = \sqrt{1 + \dfrac{1}{1}} = \sqrt{2} = 1.41421$

$x = 1.25$, $y_1 = \sqrt{1 + \dfrac{1}{1.25}} = 1.34164$

$x = 1.5$, $y_2 = \sqrt{1 + \dfrac{1}{1.5}} = 1.29099$

$x = 1.75$, $y_3 = \sqrt{1 + \dfrac{1}{1.75}} = 1.25357$

$x = 2$, $y_n = \sqrt{1 + \dfrac{1}{2}} = 1.22474$

Substituting these values into the formula gives

$$\int_1^2 \sqrt{1 + \frac{1}{x}}\, dx \approx \frac{1}{2} \times 0.25\left\{(1.41421 + 1.22474) + 2(1.34164 + 1.29099 + 1.25357)\right\}$$

≈ 1.30142

≈ 1.301 (to 3 decimal places).

Test yourself

1. Use the Trapezium rule with five ordinates to find an approximate value for the integral
$$\int_0^4 \left(\frac{1}{1 + \sqrt{x}} \right) dx$$
   Show your working and give your answer correct to three decimal places. [6]

2. Use the Trapezium rule with five ordinates to find an approximate value for the integral
$$\int_0^2 \sqrt{7 - x^2} \, dx$$
   Show your working and give your answer correct to three decimal places. [4]

3. You may assume that the equation $(x - 1)e^{2x} - 1 = 0$ has a root α between 1 and 2. The recurrence relation
$$x_{n+1} = 1 + e^{-2x_n}$$
   with $x_0 = 1.1$ can be used to find α. Find and record the values of x_1, x_2, x_3. Write down the value of x_3 correct to four decimal places and prove that this value is the value of α correct to four decimal places. [6]

4. The recurrence relation
$$x_{n+1} = (8 - 2x_n)^{\frac{1}{3}}$$
   with $x_0 = 1.7$ may be used to find the root of the equation $x^3 + 2x - 8 = 0$. Find and record the values of x_1, x_2, x_3, x_4. Write down the value of x_4 correct to four decimal places. [6]

5. Here are graphs of $y = f(x)$ and the line $y = x$ plotted on the same axes.

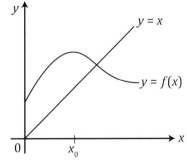

   Using x_0 as the starting value, draw a cobweb diagram to show how the sequence converges and show on your graph how the values of x_1 and x_2 can be found. [3]

6. (a) On the same diagram, sketch the graphs of $y = \ln x$ and $y = 11 - 2x$. Deduce the number of roots of the equation $\ln x + 2x - 11 = 0$. [3]

   (b) **You may assume** that the equation $\ln x + 2x - 11 = 0$ has a root α between 4 and 5.
   The recurrence relation
$$x_{n+1} = \frac{11 - \ln x_n}{2}$$
   with $x_0 = 4.7$, can be used to find α.
   Find and record the values of x_1, x_2, x_3, x_4.
   Write down the value of x_4 correct to five decimal places and prove that this is the value of α correct to five decimal places. [6]

Summary

Check you know the following facts:

Location of roots of f(x) = 0, considering changes of sign of f(x)

If $f(x)$ can take any value between a and b, then if there is a change of sign between $f(a)$ and $f(b)$, then a root of $f(x)$ lies between a and b.

Newton–Raphson iteration

Newton–Raphson iteration for solving $f(x) = 0$

$$x_{n+1} = x_n - \frac{f(x_n)}{f'(x_n)}$$

The Trapezium rule for estimating the area under a curve or the integral of a function

The Trapezium rule can be used for estimating areas or working out definite integrals of functions where the function is too difficult to integrate.

$$\int_a^b y\,dx \approx \frac{1}{2}h\{(y_0 + y_n) + 2(y_1 + y_2 + \ldots + y_{n-1})\} \quad \text{where } h = \frac{b-a}{n}$$

Test yourself answers

Topic 1

1 Start off by assuming that a is odd. If a is odd it can be written as $a = 2n + 1$, where n is any integer.

Now
$$a^2 = (2n + 1)^2$$
$$a^2 = 4n^2 + 4n + 1$$

As $4n^2 + 4n$ is even (as it has 2 as a factor), $4n^2 + 4n + 1$ must be odd so this means a^2 must be odd.

But we are given that a^2 is even, so this is a contradiction. This means that the assumption that a is odd must be wrong.

2 Hence $3n + 2n^3 = 6k + 2 \times (2k)^3 = 6k + 16k^3$

Now $6k + 16k^3$ can be factorised to give $2k(3 + 8k^2)$

$2k(3 + 8k^2)$ will always be even and as $2k(3 + 8k^2) = 3n + 2n^3$, then $3n + 2n^3$ has been proved to be even. Since $3n + 2n^3$ is odd, there is a contradiction. Hence if $3n + 2n^3$ is odd, then n is odd, so the original proposition is correct.

3 $a + b < 2\sqrt{ab}$

Squaring both sides gives
$$(a + b)^2 < 4ab$$
$$(a + b)(a + b) < 4ab$$
$$a^2 + 2ab + b^2 < 4ab$$
$$a^2 - 2ab + b^2 < 0$$
$$(a - b)^2 < 0$$

This contradicts our initial assumption, as a and b are both real numbers, so a – b must also be real.

Hence the original proposition that $a + b \geq 2\sqrt{ab}$ is true.

> Squaring both sides removes the square-root from the right-hand side of the equation.

> Note this cannot be true if a and b are both real numbers, because when squaring a real number, you always obtain a number ≥ 0.

4 Assume that 4 is a factor of $a + b$.

Then there exists an integer c such that $a + b = 4c$.

Similarly, there exists an integer d such that $a - b = 4d$.

Adding, we have $2a = 4c + 4d$.

Therefore $a = 2c + 2d$, an even number, which contradicts the fact that a is odd.

5 Assume that these is a real value of x such that
$$(5x - 3)^2 + 1 < (3x - 1)^2$$
$$25x^2 - 30x + 9 + 1 < 9x^2 - 6x + 1$$
$$16x^2 - 24x + 9 < 0$$
$$(4x - 3)^2 < 0$$

For any real value of x, $(4x - 3)^2$ is either 0 or positive so it cannot be < 0.

This contradicts the original assumption so $(5x - 3)^2 + 1 \geq (3x - 1)^2$

6. Assume that $\sqrt{5}$ is rational so that it can be expressed as $\frac{a}{b}$ where a and b are integers having no common factors.

$$\sqrt{5} = \frac{a}{b}$$

$$5 = \frac{a^2}{b^2} \text{ so } a^2 = 5b^2$$

This means a^2 has 5 as a factor.
a must also have a factor of 5 so we can write $a = 5k$, where k is an integer.
Hence, $(5k)^2 = 5b^2$ so $b^2 = 5k^2$
This means b^2 and hence b have 5 as a factor.
a and b therefore have 5 as a common factor so this is a contradiction to the original assumption meaning $\sqrt{5}$ must be irrational.

7. We assume that there is a real value of x for which $\sin x + \cos x < 1$.
Now $\sin x$ and $\cos x$ are both positive for $0 \leq x \leq \frac{\pi}{2}$ as the following graph shows:

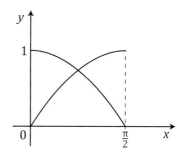

$$\sin x + \cos x < 1$$
Squaring both sides, we obtain $(\sin x + \cos x)^2 < 1^2$
$$\sin^2 x + 2\sin x \cos x + \cos^2 x < 1$$
However $\qquad\qquad \sin^2 x + \cos^2 x = 1$
So, $2\sin x \cos x + 1 < 1$, hence $2\sin x \cos x < 0$
But as both $\sin x$ and $\cos x$ are positive, $2\sin x \cos x$ cannot be less than 0 (i.e. negative) so this is a contradiction.
Hence the original statement that $\sin x + \cos x \geq 1$ is correct.

Topic 2

1. (a) The range of f is $[-1, \infty)$.
   (b) Let
   $$y = (x + 2)^2 - 1$$
   $$y + 1 = (x + 2)^2$$
   $$\pm\sqrt{y + 1} = x + 2$$
   $$-\sqrt{y + 1} - 2 = x$$
   $$f^{-1}(x) = -\sqrt{x + 1} - 2$$

   Only the negative square root is used because according to the domain $x \leq -2$.

   The domain of f^{-1} is the range of f so the domain of f^{-1} is $[-1, \infty)$.
   The range of f^{-1} is the domain of f so the range of f^{-1} is $(-\infty, -2]$.

Imagine the graph of
$$y = (x + 2)^2 - 1.$$
The curve would be U-shaped with a minimum point at $(-2, -1)$. Hence the smallest value of f is -1 at $x = -2$. According to the domain $x \leq -2$, all x values including and to the left of -2 are allowable. This will mean that the range of f will be $[-1, \infty)$.

2 $y = f(x)$ to $y = 2f(x - 1)$ represents a translation of 1 unit to the right and a stretch of scale factor 2 parallel to the y-axis.

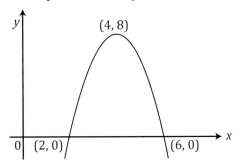

3 (a) $3|x - 1| + 7 = 19$
$3|x - 1| = 12$
$|x - 1| = 4$
$x - 1 = \pm 4$
$x = 5$ or -3

(b) $6|x| - 3 = 2|x| + 5$
$4|x| = 8$
$|x| = 2$
$x = \pm 2$

4 (a) $y = x^2 + 6x + 13$
$y = (x + 3)^2 - 9 + 13$
$y = (x + 3)^2 + 4$
Stationary point is $(-3, 4)$

> Here the method of completing the square has been used to find the coordinates of the stationary point but you could also have used differentiation to find them.

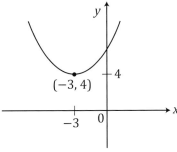

> Make sure the coordinates of the stationary point are included on the sketch.

(b) (i) f^{-1} does not exist because f is not a one-to-one function.

(ii) As any line drawn parallel to the x-axis from $x = -3$ to infinity would only cut the curve once, the domain can be any value from -3 to ∞. Any domain in this range can be given as the answer. Here we will use $(-3, \infty)$

$$y = (x + 3)^2 + 4$$
$$y - 4 = (x + 3)^2$$
$$x + 3 = \pm\sqrt{y - 4}$$
$$x = \pm\sqrt{y - 4} - 3$$
$$x = -3 \pm \sqrt{y - 4}$$

Now as x has to be -3 or larger, if the minus part of the $\pm$ were used it would result in x becoming more negative, so it would put it out of the range for the domain. Hence we use the $+$ only.

So, $x = -3 + \sqrt{y - 4}$

Hence $f^{-1}(x) = -3 + \sqrt{x - 4}$

> The function is not one-to-one because two inputs of f would correspond to a given output of f.

> Once the equation has been rearranged for x, the x is changed to $f^{-1}(x)$ and the y is replaced by x.

ln (x – 6) can only be found if x is greater than 6 as you cannot find ln of 0 or a negative number.

This means x has to be greater than 6 and can take any value from 6 to infinity.

5 (a) Let $y = e^{5-\frac{x}{2}} + 6$

$$y - 6 = e^{5-\frac{x}{2}}$$

Taking ln of both sides, we obtain $\ln(y - 6) = 5 - \frac{x}{2}$

Rearranging for x, we obtain $x = 2[5 - \ln(y - 6)]$

Hence $f^{-1}(x) = 2[5 - \ln(x - 6)]$

(b) Domain of $f^{-1} = (6, \infty)$

6 (Note that in the exam this graph would be hand-drawn.)

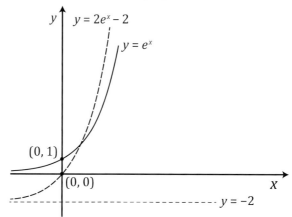

The graph of $y = f(x)$ represents a stretch of the original graph ($y = e^x$) by scale factor 2 parallel to the y-axis followed by a translation of $\begin{pmatrix} 0 \\ -2 \end{pmatrix}$.

Note these transformations can be applied the other way around.

For $y = f(x)$.
Curve cuts the y-axis at (0, 1).

7 (a) Domain of fg is the domain of $g = (0, \infty)$.
Range of fg is $(0, \infty)$.

The domain of a composite function fg is the set of all x in the domain of g for which $g(x)$ is in the domain of f.

$R(g) = (-\infty, \infty)$ and $D(f) = (-\infty, \infty)$ hence $D(fg) = D(g) = (0, \infty)$

(b) $$fg(x) = 2e^{3g(x)}$$
$$fg(x) = 2e^{3\ln 2x}$$
$$fg(x) = 2e^{\ln(2x)^3}$$
$$fg(x) = 2e^{\ln 8x^3}$$
$$fg(x) = 16x^3$$

$3\ln 2x = \ln(2x)^3$

The range of fg is found by substituting values from the domain into the composite function (i.e. $fg(x) = 16x^3$).

Since $fg(x) = 128$

$$16x^3 = 128$$
$$x^3 = 8$$
$$x = 2$$

$e^{\ln a} = a$

Topic 3

1 $$\frac{1 + x}{\sqrt{1 - 4x}} = (1 + x)(1 - 4x)^{-\frac{1}{2}}$$

Note that $1 + x$ is not involved in the condition for convergence because $1 + x$ is not an expansion.

$$= (1 + x)\left[1 + \left(-\frac{1}{2}\right)(-4x) + \frac{\left(-\frac{1}{2}\right)\left(-\frac{3}{2}\right)(-4x)^2}{2!} + ...\right]$$

$$= 1 + 2x + 6x^2 + x + 2x^2 + 6x^3 + ...$$

$$= 1 + 3x + 8x^2 + ...$$

$|4x| < 1$ so this expansion is convergent for $|x| < \frac{1}{4}$ or $-\frac{1}{4} < x < \frac{1}{4}$

② $(1 + x)^n = 1 + nx + \dfrac{n(n-1)x^2}{2!} + \ldots$

Here $n = \dfrac{1}{2}$ and x is replaced by $4x$.

$$(1 + 4x)^{\frac{1}{2}} = 1 + \left(\frac{1}{2}\right)(4x) + \frac{\left(\frac{1}{2}\right)\left(-\frac{1}{2}\right)16x^2}{1 \times 2} + \ldots$$

$$= 1 + 2x - 2x^2 + \ldots$$

Expansion is valid when $|4x| < 1$, $|x| < \dfrac{1}{4}$ or $-\dfrac{1}{4} < x < \dfrac{1}{4}$

$$(1 + 4k + 16k^2)^{\frac{1}{2}} = (1 + 4(k + 4k^2))^{\frac{1}{2}}$$

$$(1 + 4x)^{\frac{1}{2}} = 1 + 2x - 2x^2 + \ldots$$

Let $x = (k + 4k^2)$

So $(1 + 4(k + 4k^2))^{\frac{1}{2}} = 1 + 2(k + 4k^2) - 2(k + 4k^2)^2 + \ldots$

$$= 1 + 2k + 8k^2 - 2k^2 + \ldots$$

$$= 1 + 2k + 6k^2 + \ldots$$

③ $(1 + 2x)^{\frac{1}{2}} = 1 + \left(\frac{1}{2}\right)(2x) + \dfrac{\left(\frac{1}{2}\right)\left(-\frac{1}{2}\right)(2x)^2}{2!} + \dfrac{\left(\frac{1}{2}\right)\left(-\frac{1}{2}\right)\left(-\frac{3}{2}\right)(2x)^3}{3!} + \ldots$

$$= 1 + x - \frac{x^2}{2} + \frac{x^3}{2} + \ldots$$

Expansion is valid when $|2x| < 1$, so this expansion is convergent for

$$|x| < \frac{1}{2} \text{ or } -\frac{1}{2} < x < \frac{1}{2}$$

If $x = 0.01$, $(1 + 2(0.01))^{\frac{1}{2}} = (1.02)^{\frac{1}{2}} = \sqrt{1.02} \approx 1 + 0.01 - \dfrac{(0.01)^2}{2} - \dfrac{(0.01)^3}{2}$

$$\approx 1 + 0.0099505$$

$$\approx 1.009951 \text{ (correct to six decimal places)}.$$

④ $a = 4$ and $d = 6$

$S_n = \dfrac{n}{2}\big[2a + (n-1)d\big]$

$S_n = \dfrac{n}{2}\big[2 \times 4 + (n-1)6\big]$

$S_n = \dfrac{n}{2}\big(8 + 6n - 6\big)$

$S_n = \dfrac{n}{2}\big(6n + 2\big)$

$S_n = n(3n + 1)$

⑤ $S_n = \dfrac{n}{2}\big[2a + (n-1)d\big]$

$$S_7 = \frac{7}{2}\big[2a + (7-1)d\big]$$

$$182 = \frac{7}{2}\big(2a + 6d\big)$$

$$a + 3d = 26 \tag{1}$$

BOOST

Grade ⬆⬆⬆⬆

Look back at the previous result to see how it can be altered to fit the next part of the question.

Note here that this last bracket only needs expanding as far as the term in k^2.

nth term = $a + (n - 1)d$

5th term = $a + 4d$ and 7th term = $a + 6d$

$$a + 4d + a + 6d = 80$$
$$2a + 10d = 80$$

Dividing this equation by 2 gives

$$a + 5d = 40 \qquad (2)$$

Subtracting equation (1) from equation (2) gives

$$2d = 14$$
$$d = 7$$

Substituting $d = 7$ into equation (1) gives

$$26 = a + 21$$
$$a = 5$$

6 $a + ar = 2.7 \qquad (1)$

$$S_\infty = \frac{a}{1 - r}$$

$$3.6 = \frac{a}{1 - r}$$

$$a = 3.6(1 - r)$$

Substituting this into equation (1) gives

$$3.6(1 - r) + 3.6(1 - r)r = 2.7$$
$$3.6 - 3.6r + 3.6r - 3.6r^2 = 2.7$$
$$3.6r^2 = 0.9$$
$$r^2 = \frac{1}{4}$$
$$r = \pm\frac{1}{2}$$

As r must be positive, $r = \frac{1}{2}$

$$a = 3.6(1 - r)$$
$$a = 3.6\left(1 - \frac{1}{2}\right)$$
$$a = 1.8$$

7
$$t_n = a + (n - 1)d$$
$$t_{16} = a + 15d = 68 \qquad (1)$$
$$t_9 = a + 8d$$
$$t_4 = a + 3d$$

Now ninth term is double the fourth term, so

$$a + 8d = 2(a + 3d)$$
$$a = 2d$$

Substituting $a = 2d$ into equation (1) gives

$$2d + 15d = 68$$
$$17d = 68$$
$$d = 4$$
$$a = 2d = 8$$

Hence first term $a = 8$ and common difference $d = 4$.

It is best to number simultaneous equations so they can be referred to by their number.

You can write this as follows:
$$t_9 = 2t_4$$

It is easy to make a mistake when solving simultaneous equations, so remember to check them by substituting both values into the equation which has not been used for the substitution. If the right-hand side equals the left-hand side, your values are likely to be correct.

8 (a) $t_{n+1} = 2t_n + 1$

$t_4 = 2t_3 + 1$

$63 = 2t_3 + 1$

$t_3 = 31$

$t_3 = 2t_2 + 1$

$31 = 2t_2 + 1$

$t_2 = 15$

$t_2 = 2t_1 + 1$

$15 = 2t_1 + 1$

$t_1 = 7$

(b) 6 043 582 is even but all the terms of the sequence are odd.

2 × (an even or odd number) always results in an even number and adding a one to an even number will make an odd number.

9 $6\sqrt{1 - 2x} - \dfrac{1}{1 + 4x} = 6(1 - 2x)^{\frac{1}{2}} - (1 + 4x)^{-1}$

> This formula is obtained from the formula booklet.

$(1 + x)^n = 1 + nx + \dfrac{n(n - 1)x^2}{2!} + \dots$

$(1 - 2x)^{\frac{1}{2}} = 1 + \left(\dfrac{1}{2}\right)(-2x) + \dfrac{\left(\frac{1}{2}\right)\left(-\frac{1}{2}\right)(-2x)^2}{2} + \dots = 1 - x - \dfrac{x^2}{2} + \dots$

$(1 + 4x)^{-1} = 1 + (-1)(4x) + \dfrac{(-1)(-2)(4x)^2}{2} + \dots = 1 - 4x + 16x^2 + \dots$

$6(1 - 2x)^{\frac{1}{2}} - (1 + 4x)^{-1} = 6\left(1 - x - \dfrac{x^2}{2}\right) - (1 - 4x + 16x^2)$

$= 6 - 6x - 3x^2 - 1 + 4x - 16x^2$

$= 5 - 2x - 19x^2$

$(1 - 2x)^{\frac{1}{2}}$ is valid for $\left|\dfrac{2x}{1}\right| < 1$ so $|x| < \dfrac{1}{2}$

$(1 + 4x)^{-1}$ is valid for $\left|\dfrac{4x}{1}\right| < 1$ so $|x| < \dfrac{1}{4}$

Now as the value of x has to satisfy both conditions

and as $|x| < \dfrac{1}{4}$ lies inside $|x| < \dfrac{1}{2}$, the condition for the combined series is $|x| < \dfrac{1}{4}$.

BOOST

Grade ⬆⬆⬆⬆

Remember that for a series that is composed of two series you need to find the conditions for validity for each series and see if one lies inside the other.

Topic 4

1 (a) $3\cos\theta + 2\sin\theta = R\cos(\theta - \alpha)$

$3\cos\theta + 2\sin\theta = R\cos\theta\cos\alpha + R\sin\theta\sin\alpha$

$R\cos\alpha = 3$ and $R\sin\alpha = 2$

$\tan\alpha = \dfrac{2}{3}$ so $\alpha = 33.7°$

$R = \sqrt{3^2 + 2^2} = \sqrt{13}$

Hence, $3\cos\theta + 2\sin\theta = \sqrt{13}\cos(\theta - 33.7°)$

(b) $\sqrt{13} \cos (\theta - 33.7°) = 1$

$$\cos (\theta - 33.7°) = \frac{1}{\sqrt{13}}$$

$$\theta - 33.7° = 73.9°, 286.1°$$
$$\theta = 107.6°, 319.8°$$

Use $\cos 2\theta = 1 - 2\sin^2 \theta$ to obtain a quadratic equation in just $\sin \theta$.

2
$$3 \cos 2\theta = 1 - \sin \theta$$
$$3(1 - 2\sin^2 \theta) = 1 - \sin \theta$$
$$3 - 6\sin^2 \theta = 1 - \sin \theta$$
$$6 \sin^2 \theta - \sin \theta - 2 = 0$$
$$(3 \sin \theta - 2)(2 \sin \theta + 1) = 0$$

Hence $\sin \theta = \dfrac{2}{3}$ or $\sin \theta = -\dfrac{1}{2}$

When $\sin \theta = \dfrac{2}{3}, \theta = 41.8°, 138.2°$

When $\sin \theta = -\dfrac{1}{2}, \theta = 210°, 330°$

Hence $\theta = 41.8°, 138.2°, 210°, 330°$

3 $4 \sin \theta + 5 \cos \theta \equiv R \cos(\theta - \alpha)$
$4 \sin \theta + 5 \cos \theta \equiv R \cos \theta \cos \alpha + R \sin \theta \sin \alpha$
$$R \cos \alpha = 5 \text{ and } R \sin \alpha = 4$$
$$\tan \alpha = \frac{4}{5} \text{ so } \alpha = 38.7°$$
$$R = \sqrt{4^2 + 5^2} = \sqrt{41}$$

Hence, $\quad 4 \sin \theta + 5 \cos \theta = \sqrt{41} \cos (\theta - 38.7°)$
$$\sqrt{41} \cos (\theta - 38.7°) = 2$$

giving $\quad \cos (\theta - 38.7°) = \dfrac{2}{\sqrt{41}}$

$$(\theta - 38.7°) = 71.8°, 288.2°$$
Hence $\theta = 110.5°, 326.9°$

4 Let $\theta = \dfrac{\pi}{2}$

LHS $= \cos 4\theta = \cos 4\left(\dfrac{\pi}{2}\right) = \cos 2\pi = 1$

Note that $\quad \cos \dfrac{\pi}{2} = 0$
and $\quad\quad \cos 2\pi = 1$

RHS $= 4 \cos^3 \theta - 3 \cos \theta = 4 \cos^3 \dfrac{\pi}{2} - 3 \cos \dfrac{\pi}{2} = 0$

$1 \neq 0$ so the statement $\cos 4\theta \equiv 4 \cos^3 \theta - 3 \cos \theta$ is false.

You need to use the trigonometric identity
$$\sec^2 \theta = 1 + \tan^2 \theta.$$
This identity must be remembered.

5 $2 \sec^2 \theta + \tan \theta = 8$
$$2(1 + \tan^2 \theta) + \tan \theta - 8 = 0$$
$$2 \tan^2 \theta + \tan \theta - 6 = 0$$
$$(2 \tan \theta - 3)(\tan \theta + 2) = 0$$

$\tan \theta$ is positive in the first and third quadrants and negative in the second and fourth quadrants.

$\tan \theta = \dfrac{3}{2}$ or $\tan \theta = -2$

When $\tan \theta = \dfrac{3}{2}, \theta = 56.3°$ or $236.3°$

When $\tan \theta = -2, \theta = 180 - 63.4 = 116.6°$ or $\theta = 360 - 63.4 = 296.6°$
Hence solutions are $\theta = 56.3°, 116.6°, 236.3°$ or $296.6°$

6 (a) Let $\theta = \dfrac{\pi}{6}$

LHS $= \tan 2\theta = \tan \dfrac{\pi}{3} = \sqrt{3}$

RHS $= \dfrac{2 \tan \theta}{1 + \tan^2 \theta} = \dfrac{2 \tan \dfrac{\pi}{6}}{1 + \tan^2 \dfrac{\pi}{6}} = \dfrac{\dfrac{2}{\sqrt{3}}}{1 + \dfrac{1}{3}} = \dfrac{3}{2\sqrt{3}}$

Note $\tan \dfrac{\pi}{6} = \dfrac{1}{\sqrt{3}}$

$\sqrt{3} \neq \dfrac{3}{2\sqrt{3}}$ so the statement $\tan 2\theta \equiv \dfrac{2 \tan \theta}{1 + \tan^2 \theta}$ is false.

(b)
$$2 \sec \theta + \tan^2 \theta = 7$$
$$2 \sec \theta + \sec^2 \theta - 1 = 7$$
$$\sec^2 \theta + 2 \sec \theta - 8 = 0$$
$$(\sec \theta - 2)(\sec \theta + 4) = 0$$
Hence $\sec \theta = 2$ or $\sec \theta = -4$

So $\dfrac{1}{\cos \theta} = 2$ or $\dfrac{1}{\cos \theta} = -4$

Hence $\cos \theta = \dfrac{1}{2}$ or $\cos \theta = -\dfrac{1}{4}$

When $\cos \theta = \dfrac{1}{2}$, $\theta = 60°$ or $300°$

When $\cos \theta = -\dfrac{1}{4}$, $\theta = 104.5°$ or $255.5°$

Hence $\theta = 60°, 104.5°, 255.5°$ or $300°$

$\cos \theta$ is negative in the second and third quadrants, so $\theta = 180 - 75.5 = 104.5°$ or $360 - 75.5 = 255.5°$

7 Area of sector AOB $= \dfrac{1}{2} r^2 \theta = \dfrac{1}{2} r^2 (2.6) = 1.3 r^2$

Area of triangle AOB $= \dfrac{1}{2} r^2 \sin(2.6) = 0.2578 r^2$

Note that the formula Area of triangle $= \dfrac{1}{2} ab \sin C$ has been used here to work out the area of the triangle.

Area of minor segment $= 1.3 r^2 - 0.2578 r^2 = 1.0422 r^2$
Area of major segment $=$ Area of circle $-$ Area of minor segment
$$= \pi r^2 - 1.0422 r^2$$
$$= 2.0994 r^2$$
Now, Area of major segment $\approx 2 \times$ Area of minor segment.
Hence area of segment for white roses is approximately twice the area containing red roses.

Topic 5

1 $y = (4x^3 + 3x)^3$

$$\dfrac{dy}{dx} = 3(4x^3 + 3x)^2 (12x^2 + 3)$$

$$= 9(4x^2 + 1)(4x^3 + 3x)^2$$

2 $y = (3 - 2x)^{10}$

$$\dfrac{dy}{dx} = 10(3 - 2x)^9 (-2)$$

$$= -20(3 - 2x)^9$$

3 $x = 3t^2$

$$\frac{dx}{dt} = 6t$$

$y = t^4$

$$\frac{dy}{dt} = 4t^3$$

$$\frac{dy}{dx} = \frac{dy}{dt} \times \frac{dt}{dx}$$

$$= 4t^3 \times \left(\frac{1}{6t}\right)$$

$$= \frac{2}{3}t^2$$

4 $4x^3 - 6x^2 + 3xy = 5$

Differentiating implicitly with respect to x gives

$$12x^2 - 12x + (3x)(1)\left(\frac{dy}{dx}\right) + (y)(3) = 0$$

$$12x^2 - 12x + (3x)\left(\frac{dy}{dx}\right) + 3y = 0$$

Simplify by dividing both sides by 3.

$$3x\left(\frac{dy}{dx}\right) = 12x - 12x^2 - 3y$$

$$x\left(\frac{dy}{dx}\right) = 4x - 4x^2 - y$$

$$\frac{dy}{dx} = \frac{4x - 4x^2 - y}{x}$$

5 (a) $y = \ln(x^3)$

$$\frac{dy}{dx} = \frac{3x^2}{x^3}$$

$$= \frac{3}{x}$$

Note also that $\ln x^3 = 3 \ln x$ (laws of logs) so we are in effect differentiating $3 \ln x$.

(b) $y = \ln(\sin x)$

$$\frac{dy}{dx} = \frac{\cos x}{\sin x} = \cot x$$

6 (a) $y = (2x^2 - 1)^3$

$$\frac{dy}{dx} = 3(2x^2 - 1)^2(4x)$$

$$= 12x(2x^2 - 1)^2$$

Use the Chain rule with $u = 2x^2 - 1$, or use the table on page 113 with $f(x) = 2x^2 - 1, n = 3$.

(b) $y = x^3 \sin 2x$

$$\frac{dy}{dx} = x^3 \, 2 \cos 2x + \sin 2x(3x^2)$$

$$= 2x^3 \cos 2x + 3x^2 \sin 2x$$

$$= x^2(2x \cos 2x + 3 \sin 2x)$$

This is the product of two functions, so the Product rule must be used when differentiating.

(c) $y = \dfrac{3x^2 + 4}{x^2 + 6}$

$$\frac{dy}{dx} = \frac{(6x)(x^2+6) - (3x^2+4)(2x)}{(x^2+6)^2}$$

$$= \frac{6x^3 + 36x - 6x^3 - 8x}{(x^2+6)^2}$$

$$= \frac{28x}{(x^2+6)^2}$$

7 Differentiating implicitly with respect to x, we obtain

$$3x^2 + (6x)(2y)\left(\frac{dy}{dx}\right) + y^2(6) = 3y^2\frac{dy}{dx}$$

$$3x^2 + 12xy\left(\frac{dy}{dx}\right) + 6y^2 = 3y^2\frac{dy}{dx}$$

$$(12xy - 3y^2)\left(\frac{dy}{dx}\right) = -3x^2 - 6y^2$$

$$\frac{dy}{dx} = \frac{-3x^2 - 6y^2}{12xy - 3y^2} = \frac{-3(x^2+2y^2)}{3y(4x-y)} = \frac{-(x^2+2y^2)}{y(4x-y)}$$

The term $6xy^2$ is differentiated using the Product rule.

8 $y = \ln(2 + 5x^2)$

$$\frac{dy}{dx} = \frac{10x}{(2+5x^2)}$$

Topic 6

1 $\dfrac{dx}{dt} = 6t$ and $\dfrac{dy}{dt} = 3t^2$

Hence $\dfrac{dy}{dx} = \dfrac{dy}{dt} \times \dfrac{dt}{dx} = 3t^2 \times \dfrac{1}{6t} = \dfrac{t}{2}$

Gradient of normal $= -\dfrac{2}{t}$

At P $(3p^2, p^3)$ gradient of normal $= -\dfrac{2}{p}$

Equation of the normal is $\quad y - p^3 = -\dfrac{2}{p}(x - 3p^2)$

$$py - p^4 = -2x + 6p^2$$
$$py + 2x = 6p^2 + p^4$$
$$py + 2x = p^2(6 + p^2)$$

2 $y^2 - 5xy + 8x^2 = 2$

Differentiating with respect to x gives

$$2y\frac{dy}{dx} - (5x)\frac{dy}{dx} + (y)(-5) + 16x = 0$$

$$\frac{dy}{dx}(2y - 5x) = 5y - 16x$$

Hence $\quad \dfrac{dy}{dx} = \dfrac{5y - 16x}{2y - 5x}$

Test yourself answers

3 (a) $\frac{dx}{dt} = 8\cos 4t,\quad \frac{dy}{dt} = -4\sin 4t$

$$\frac{dy}{dx} = \frac{dy}{dt} \times \frac{dt}{dx}$$

$$= -4\sin 4t \times \frac{1}{8\cos 4t}$$

$$= \frac{-\sin 4t}{2\cos 4t}$$

$$= -\frac{1}{2}\tan 4t$$

(b) $y - y_1 = m(x - x_1)$

$$y - \cos 4p = -\frac{\sin 4p}{2\cos 4p}(x - 2\sin 4p)$$

$$2y\cos 4p - 2\cos^2 4p = -x\sin 4p + 2\sin^2 4p$$
$$2y\cos 4p + x\sin 4p = 2\sin^2 4p + 2\cos^2 4p$$
$$2y\cos 4p + x\sin 4p = 2(\sin^2 4p + \cos^2 4p)$$
$$2y\cos 4p + x\sin 4p = 2$$

4 (a) $\frac{dx}{dt} = 2t \quad$ and $\quad \frac{dy}{dt} = 3t^2$

Hence $\quad \frac{dy}{dx} = \frac{dy}{dt} \times \frac{dt}{dx} = 3t^2 \times \frac{1}{2t} = \frac{3}{2}t$

At P (p^2, p^3), $\quad \frac{dy}{dx} = \frac{3}{2}p$

(b) Equation of the tangent at P is $\quad y - p^3 = \frac{3}{2}p(x - p^2)$

$$2y - 2p^3 = 3px - 3p^3$$
$$3px - 2y = p^3$$

5 (a) $4x^2 - 6xy + y^2 = 20$

Differentiating with respect to x gives

$$8x - 6x\frac{dy}{dx} - y(6) + 2y\frac{dy}{dx} = 0$$

$$(2y - 6x)\frac{dy}{dx} = 6y - 8x$$

$$\frac{dy}{dx} = \frac{6y - 8x}{2y - 6x}$$

$$= \frac{3y - 4x}{y - 3x}$$

(b) $4x^2 - 6xy + y^2 = 20$

Substituting $x = 0$ into this equation gives
$$4(0)^2 - 6(0)y + y^2 = 20$$
Hence $\quad\quad\quad y^2 = 20$

$$y = \pm\sqrt{20}$$
$$y = \pm\sqrt{4 \times 5}$$
$$y = \pm 2\sqrt{5}$$

6 (a) For toy car A, $x = 40t - 40$ and $y = 120t - 160$
$3x = 120t - 120$ so $120t = 3x + 120$
Combining the two equations, we obtain $y = 3x + 120 - 160$
$$y = 3x - 40$$

This is the Cartesian equation of car A.

For toy car B, $x = 30t$, $y = 20t^2$
$t = \dfrac{x}{30}$ so $y = 20t^2 = 20\left(\dfrac{x}{30}\right)^2$

Hence $\qquad y = \dfrac{20x^2}{900}$

$$y = \dfrac{x^2}{45}$$
$$45y = x^2$$

This is the Cartesian equation of car B.

Solving the two Cartesian equations simultaneously we have:
$$y = 3x - 40$$
$$45y = 135x - 1800$$
But $45y = x^2$, so $x^2 = 135x - 1800$
Hence $x^2 - 135x + 1800 = 0$
$(x - 120)(x - 15) = 0$
$x = 120$ or 15
When $x = 120$, $y = 320$ and when $x = 15$, $y = 5$
So paths intersect at $(120, 320)$ and $(15, 5)$

This is a difficult quadratic to solve by factorisation. In cases like this it may be quicker to use the formula.

(b) For toy car A, $x = 40t - 40$, so $120 = 40t - 40$ giving $t = 4$ s
$$x = 40t - 40, \text{ so } 15 = 40t - 40 \text{ giving } t = \dfrac{55}{40} = \dfrac{11}{8}\text{ s}$$

For toy car B, $x = 30t$, so $120 = 30t$ giving $t = 4$ s
$$x = 30t, \text{ so } 15 = 30t \text{ giving } t = \dfrac{1}{2}\text{ s}$$

As cars A and B at $t = 4$ s are at the same place (i.e. $(120, 320)$), the cars will collide.
They will not collide at the other point as they are not there at the same time.

Topic 7

1 (a) $\displaystyle\int \dfrac{6}{5x + 1}\,dx = 6\int \dfrac{1}{5x + 1}\,dx = \dfrac{6}{5}\ln|5x + 1| + c$

(b) $\displaystyle\int \cos 7x\,dx = \dfrac{1}{7}\sin 7x + c$

(c) $\displaystyle\int \dfrac{4}{(3x + 1)^3}\,dx = 4\int \dfrac{1}{(3x + 1)^3}\,dx$

$\qquad\qquad = 4\int (3x + 1)^{-3}\,dx$

$\qquad\qquad = \dfrac{4}{3 \times (-2)}(3x + 1)^{-2} + c = -\dfrac{2}{3}(3x + 1)^{-2} + c$

2 (a) $\int \sin 4x \, dx = -\dfrac{1}{4}\cos 4x + c$

(b) $\int \dfrac{1}{2x + 1} \, dx = \dfrac{1}{2}\ln|2x + 1| + c$

(c) $\int \dfrac{4}{(2x + 1)^5} \, dx = 4\int (2x + 1)^{-5} \, dx$

$$= -\dfrac{1}{2}(2x + 1)^{-4} + c$$

3 $\displaystyle\int_0^2 \dfrac{1}{(2x + 1)^3} \, dx = \int_0^2 (2x + 1)^{-3} \, dx$

$$= \left[-\dfrac{1}{4}(2x + 1)^{-2} \right]_0^2$$

$$= -\dfrac{1}{4}\left[\dfrac{1}{(2x + 1)^2} \right]_0^2$$

$$= -\dfrac{1}{4}\left[\left(\dfrac{1}{25}\right) - (1) \right]$$

$$= \left(-\dfrac{1}{4}\right) \times \left(-\dfrac{24}{25}\right)$$

$$= \dfrac{6}{25}$$

4 $\displaystyle\int_0^3 \dfrac{1}{5x + 2} \, dx = \dfrac{1}{5}\Big[\ln|5x + 2|\Big]_0^3$

$$= \dfrac{1}{5}\Big[\ln 17 - \ln 2\Big]$$

$$= \dfrac{1}{5}\ln\left(\dfrac{17}{2}\right)$$

5 $\int u\dfrac{dv}{dx} \, dx = uv - \int v\dfrac{du}{dx} \, dx$

Use Rule 1.

Let $u = 2x + 1$ and $\dfrac{dv}{dx} = e^{2x}$

So $\dfrac{du}{dx} = 2$, $v = \dfrac{e^{2x}}{2}$

$$\int (2x + 1)e^{2x} \, dx = (2x + 1)\dfrac{1}{2}e^{2x} - \int \dfrac{1}{2}e^{2x}(2) \, dx$$

$$= \dfrac{1}{2}e^{2x}(2x + 1) - \dfrac{1}{2}e^{2x} + c$$

$$= xe^{2x} + c$$

6 Let $x = 2 \sin \theta$, so $\dfrac{dx}{d\theta} = 2 \cos \theta$ giving $dx = 2 \cos \theta \, d\theta$

When $x = 1$, $1 = 2 \sin \theta$, hence $\theta = \sin^{-1}\left(\dfrac{1}{2}\right) = \left(\dfrac{\pi}{6}\right)$

When $x = 0$, $0 = 2 \sin \theta$, hence $\theta = \sin^{-1} 0 = 0$

$$\int_0^1 \sqrt{(4 - x^2)} \, dx = \int_0^{\frac{\pi}{6}} \sqrt{(4 - 4 \sin^2 \theta)} \, (2 \cos \theta) \, d\theta$$

$$= \int_0^{\frac{\pi}{6}} \sqrt{4(1 - \sin^2 \theta)} \, (2 \cos \theta) \, d\theta$$

$$= \int_0^{\frac{\pi}{6}} \sqrt{4 \cos^2 \theta} \, (2 \cos \theta) \, d\theta$$

$$= \int_0^{\frac{\pi}{6}} 2 \cos \theta \, (2 \cos \theta) \, d\theta$$

$$= \int_0^{\frac{\pi}{6}} 4 \cos^2 \theta \, d\theta$$

$$= \int_0^{\frac{\pi}{6}} 4 \left(\dfrac{1 + \cos 2\theta}{2} \right) d\theta$$

$$= 2 \int_0^{\frac{\pi}{6}} (1 + \cos 2\theta) \, d\theta$$

$$= 2 \left[\theta + \dfrac{1}{2} \sin 2\theta \right]_0^{\frac{\pi}{6}}$$

$$= 2 \left[\left(\dfrac{\pi}{6} + \dfrac{1}{2} \sin \dfrac{\pi}{3} \right) - \left(0 + \dfrac{1}{2} \sin 0 \right) \right]$$

$$= 1.913 \text{ (correct to three decimal places).}$$

> $1 - \sin^2 \theta = \cos^2 \theta$.

> This is a rearrangement of the double angle formula
> $\cos 2A = 2 \cos^2 A - 1$.

7 $\dfrac{dy}{dx} = \dfrac{y}{x + 2}$

Separating variables and integrating, we obtain

$$\int \dfrac{1}{y} dy = \int \dfrac{1}{x + 2} dx$$

So that $\qquad \ln y = \ln (x + 2) + c$

When $x = 0, y = 2$,

$\therefore \qquad\qquad\qquad \ln 2 = \ln 2 + c$

$\therefore \qquad\qquad\qquad\qquad c = 0$

and the solution is $\qquad \ln y = \ln (x + 2)$

Take exponentials to remove logs

$\therefore \qquad\qquad\qquad\qquad y = x + 2$

Topic 8

1 $\int_a^b y\,dx \approx \frac{1}{2}h\{(y_0 + y_n) + 2(y_1 + y_2 + \ldots + y_{n-1})\}$

$\int_0^4 \frac{1}{1 + \sqrt{x}}\,dx \approx \frac{1}{2}h\{(y_0 + y_n) + 2(y_1 + y_2 + \ldots + y_{n-1})\}$

$h = \dfrac{b - a}{n} = \dfrac{4 - 0}{4} = 1$

When

$x = 0,\qquad y_0 = \dfrac{1}{1 + \sqrt{0}} = 1$

$x = 1,\qquad y_1 = \dfrac{1}{1 + \sqrt{1}} = 0.5$

$x = 2,\qquad y_2 = \dfrac{1}{1 + \sqrt{2}} = 0.41421$

$x = 3,\qquad y_3 = \dfrac{1}{1 + \sqrt{3}} = 0.36603$

$x = 4,\qquad y_4 = \dfrac{1}{1 + \sqrt{4}} = 0.33333$

$\int_0^4 \sqrt{\dfrac{1}{1 + \sqrt{x}}}\,dx \approx \frac{1}{2} \times 1\{(1 + 0.33333) + 2(0.5 + 0.41421 + 0.36603)\}$

$\approx 1.94691 \approx 1.947$ to 3 decimal places.

> *n* is the number of strips and not the number of ordinates. Here the number of strips is 4.

2 $h = \dfrac{b - a}{n} = \dfrac{2 - 0}{4} = 0.5$

When $\quad x = 0,\qquad y_0 = \sqrt{7 - 0} = \sqrt{7} = 2.6458$

$x = 0.5,\qquad y_1 = \sqrt{7 - 0.5^2} = 2.5981$

$x = 1.0,\qquad y_2 = \sqrt{7 - 1^2} = 2.4495$

$x = 1.5,\qquad y_3 = \sqrt{7 - 1.5^2} = 2.1794$

$x = 2.0,\qquad y_4 = \sqrt{7 - 2^2} = 1.7321$

$\int_a^b y\,dx \approx \frac{1}{2}h\{(y_0 + y_n) + 2(y_1 + y_2 + \ldots y_{n-1})\}$

$\int_0^2 \sqrt{7 - x^2}\,dx \approx \frac{1}{2} \times 0.5\{(2.6458 + 1.7321) + 2(2.5981 + 2.4495 + 2.1794)\}$

≈ 4.7080

≈ 4.708 (to 3 decimal places)

3 $x_0 = 1.1$

$x_1 = 1 + e^{-2(1.1)} = 1.11080316$

$x_2 = 1 + e^{-2(1.11080316)} = 1.10843479$

$x_3 = 1 + e^{-2(1.10843479)} = 1.10894963$

$x_3 = 1.1089$ (correct to 4 decimal places)

Let $f(x) = (x - 1)e^{2x} - 1$

$$f(1.10885) = (1.10885 - 1)e^{2(1.10885)} - 1 = -0.000084$$
$$f(1.10895) = (1.10895 - 1)e^{2(1.10895)} - 1 = 0.001034$$

As there is a sign change, $\alpha = 1.1089$ correct to four decimal places.

4 $x_0 = 1.7$

$$x_1 = (8 - 2x_0)^{\frac{1}{3}} = (8 - 2(1.7))^{\frac{1}{3}} = 1.663103499$$

$$x_2 = (8 - 2x_1)^{\frac{1}{3}} = (8 - 2(1.663103499))^{\frac{1}{3}} = 1.671949509$$

$$x_3 = (8 - 2x_2)^{\frac{1}{3}} = (8 - 2(1.671949509))^{\frac{1}{3}} = 1.669837194$$

$$x_4 = (8 - 2x_3)^{\frac{1}{3}} = (8 - 2(1.669837194))^{\frac{1}{3}} = 1.670342073$$

Hence $x_4 = 1.6703$ (correct to 4 decimal places)

Always round your final answer to the required number of decimal places or significant figures.

5

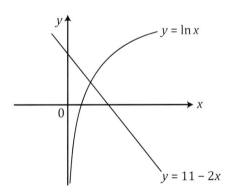

6 (a)

Notice that there is one point of intersection so there is one root of the equation $\ln x + 2x - 11 = 0$.

(b) $x_0 = 4.7$

The values of x_1 to x_4 are as follows:

$$x_1 = 4.726218746$$
$$x_2 = 4.723437268$$
$$x_3 = 4.723731615$$
$$x_4 = 4.723700458$$

Hence $x_4 = 4.72370$ to 5 decimal places.

Checking the sign of $h(x)$ at $x = 4.723695$ gives $h(x) = -1.87 \times 10^{-5}$

Checking the sign of $h(x)$ at $x = 4.723705$ gives $h(x) = 3.45 \times 10^{-6}$

As there is a sign change between these two points

$\alpha = 4.72370$ to 5 decimal places.